Socially Engaged Art and the Neoliberal City

What are the social functions of art in the age of neoliberal urbanism? This book discusses the potential of artistic practices to question the nature of city environments and the diverse productions of space, moving beyond the reduction of 'the urban' as a set of existing and static structures. Adopting a practice-led approach, each chapter discusses case studies from across the world, reflecting on personal experiences as well as the work of other artists. While exposing the increasingly limiting constraints placed on public and socially engaged art by the dominance of commercial funding and neoliberal frameworks, the author stays optimistic about the potential of artistic practices to transcend neoliberal logics through alternative productions of space. Drawing upon a Lefebvrian framework of spatial practice and using a structuralist approach to challenge neoliberal structures, the book draws links between art, resistance, criticism, democracy, and political change.

The book concludes by looking at how we might create a new course for socially engaged art within the neoliberal city. It will be of great interest to researchers in urban studies, urban geography, and architecture, as well as students who want to learn more about place-making, visual culture, performance theory, applied practice, and urban culture.

Cecilie Sachs Olsen is a British Academy Post Doctoral Researcher at Royal Holloway University of London, UK.

Routledge Critical Studies in Urbanism and the City

This series offers a forum for cutting-edge and original research that explores different aspects of the city. Titles within this series critically engage with, question and challenge contemporary theory and concepts to extend current debates and pave the way for new critical perspectives on the city. This series explores a range of social, political, economic, cultural and spatial concepts, offering innovative and vibrant contributions, international perspectives and interdisciplinary engagements with the city from across the social sciences and humanities.

Urban Subversion and the Creative City
Oli Mould

Mega-Event Mobilities
A Critical Analysis
Edited by Noel B. Salazar, Christiane Timmerman, Johan Wets, Luana Gama Gato and Sarah Van den Broucke

Art and the City
Worlding the Discussion through a Critical Artscape
Edited by Julie Ren and Jason Luger

Gentrification as a Global Strategy
Neil Smith and Beyond
Edited by Abel Albet and Núria Benach

Gender and Gentrification
Winifred Curran

Socially Engaged Art and the Neoliberal City
Cecilie Sachs Olsen

For more information about this series, please visit www.routledge.com/Routledge-Critical-Studies-in-Urbanism-and-the-City/book-series/RSCUC

Socially Engaged Art and the Neoliberal City

Cecilie Sachs Olsen

Routledge
Taylor & Francis Group
LONDON AND NEW YORK

First published 2019
by Routledge
2 Park Square, Milton Park, Abingdon, Oxon OX14 4RN

and by Routledge
52 Vanderbilt Avenue, New York, NY 10017

Routledge is an imprint of the Taylor & Francis Group, an informa business

British Library Cataloguing-in-Publication Data
A catalogue record for this book is available from the British Library

Library of Congress Cataloging-in-Publication Data
Names: Olsen, Cecilie Sachs, author.
Title: Socially engaged art and the neoliberal city / Cecilie Sachs Olsen.
Description: First edition. | New York : Routledge, [2019] | Series: Routledge critical studies in urbanism and the city | Includes index.
Identifiers: LCCN 2018043434| ISBN 9781138343399 (hbk : alk. paper) | ISBN 9780429439162 (ebk) | ISBN 9780429799150 (mobi/kindle)
Subjects: LCSH: Artists and community. | Art and society. | Sociology, Urban.
Classification: LCC N72.A76 O47 2019 | DDC 700.1/03—dc23
LC record available at https://lccn.loc.gov/2018043434

ISBN: 978-1-138-34339-9 (hbk)
ISBN: 978-0-429-43916-2 (ebk)

Typeset in Times New Roman
by Swales & Willis Ltd, Exeter, Devon, UK

MIX
Paper from responsible sources
FSC® C013056

Printed and bound in Great Britain by
TJ International Ltd, Padstow, Cornwall

For Vilja, the world is yours to make.

Contents

Figures

Acknowledgements

This book is the result of countless encounters, collaborative efforts, and experiments, and as such owes much in its form and content to the many artists and practitioners with whom I have been fortunate to share my research and practice. This is a brief list of the people to whom I am grateful in this regard.

First of all, I would like to thank Sabeth Tödtli and Nina Lund Westerdahl, the core team of my artist collective, zURBS, who have accommodated my research on our work in such a generous way. Their critical questioning, creative inputs, and unconditional support throughout the process were invaluable.

A big thank you goes to all the people who helped us organize and execute the artistic projects I discuss in my research, in particular Sander van Parijs, Laura Bruns, Lara Guth, Markus Nollert, J. B. Beovardi, Christoph Goetschi, Agnes Jones, Peter Moreton, Alison Turnbull, Imanuel Schipper, Johanna Dangel, Nico Grüninger, Fabien Thétaz, Melisende Navarre, Roger Merguin, and Kathrin Veser. And thank you to all the participants of our workshops who shared their thoughts and experiences with me. I hope to have been able, in my interpretations and narratives, to honour some of the richness of their knowledge and understandings.

For their brilliant guidance, I would like to thank David Pinder, Jen Harvie, and Harriet Hawkins, who have offered a deep understanding of my work, a constant encouragement, and in-depth critical commentary. I am also indebted to the stimulating environment of the School of Geography and the School of English and Drama at Queen Mary University of London, the Department of Geography at Royal Holloway University of London, as well as the inspiring world of art-geography scholarship in London and internationally.

My reflections on the intersection between art and geography would not have been possible without the generous support of, and discussions with, scholars such as Jens Badura, Steve Pile, Jane Rendell, and Alan Read. I am also grateful to the artists who have shared their practical and artistic reflections with me, in particular Richard Wentworth, Jeanne van Heeswijk, and Jill Sigman.

I also want to thank the following for granting permission to reproduce the images: Richard Wentworth, DACS, and the Lisson Gallery, Rotor, Jill Sigman and Rafael Gamo, Group Material and Doug Ashford, Jeanne van Heeswijk, and

Stadt Zürich. Thanks must also go to Ruth Anderson and Faye Leerink for their editorial help. And finally, I wish to thank the British Academy for their generous funding, which has made it possible for me to write this book.

The content in Chapters 1 and 2 is based on Sachs Olsen, C. "Urban Space and the Politics of Socially Engaged Art." *Progress in Human Geography*, 2018.

The content in Chapters 7, 8 and 9 is based on Sachs Olsen, C. "Collaborative Challenges: Negotiating the complicities of socially engaged art within an era of neoliberal urbanism." *Environment and Planning D: Society and Space*, 2018.

Lastly, I would never have been able to bring this research to completion without the unfailing support of my family and friends. Thank you all for believing in me, for participating in workshops and for being such dedicated fans of zURBS. A special and final thanks to Matt, for always challenging me in the best ways possible.

Introduction

Producing urban imaginaries in socially engaged art

You participate in a workshop that is part of the socially engaged art project invisible Zürichs. *Having spent the last hour searching the city streets with your fellow participants for traces of the many different 'invisible' stories that make up Zurich, you are now returning to 'the alternative city archive' with your findings: a well-used ironing board neatly left behind on the sidewalk in the Jewish quarter; a crutch haphazardly discarded in a bush; a half-full glass of red wine forgotten at a sticky restaurant table; a shopping receipt that reveals clear intentions by listing dinner ingredients and lubricant; a glass splinter from a broken window in the red-light district; an empty bottle of brandy placed on a staircase in the business district; and numerous gloves, hats, umbrella covers, and Christmas decorations that are proof of a cold November and rainy December. The objects tell of a diverse set of experiences and stories pointing to different understandings, perspectives, and perceptions of Zurich: the glass splinter is seen as an example of the air in 'the frozen state' Zurich, commenting upon the perceived conservatism of the city; the shopping receipt becomes a popular conversation piece on dating and isolation among singles in the city; and the various alcoholic beverages stemming from the newly built business district become part of an ongoing critique of this area as purely one of 'work and party' for expats and businessmen.*

You notice that these stories are not merely recounting events and descriptions of Zurich, but that they endow the city with meaning by commentary, interpretation, and dramatic structure. One of the other participants sees the lonely wine glass as evidence of transgression and tells a story about the courageous person that goes to a restaurant to eat her dinner all alone among families and couples. Another participant hangs the ironing board from the ceiling and proclaims that it is now a 'conversation bridge' that can be pushed between people having heated discussions, in order to release tension and have a calm conversation. The different stories make you think about how you want to live together and how you would like your city to be: Would you feel weird having a glass of wine alone in a restaurant? Is it the 'work and party' atmosphere of the business district that makes you not feel welcome there? What is it that makes Zurich feel like a frozen state?

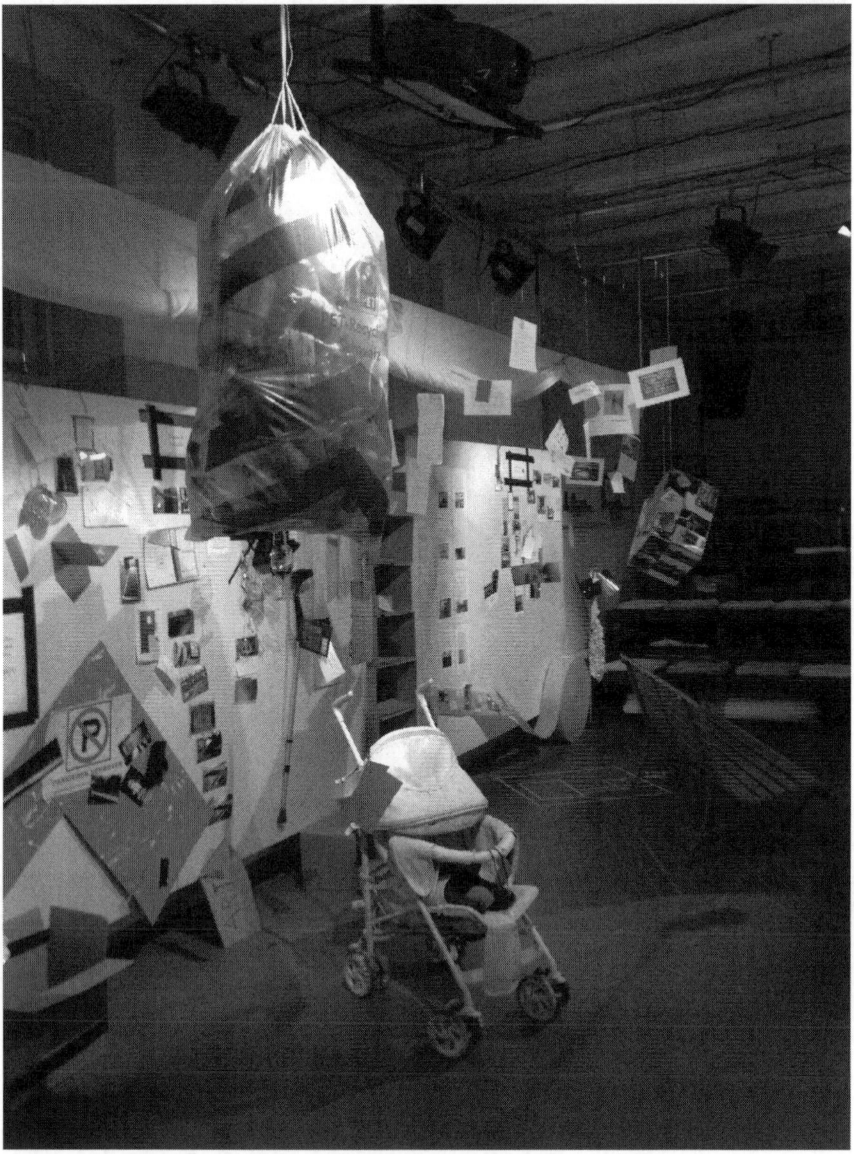

Figure I.1 'The alternative city archive', *invisible Zürichs*, zURBS, 2013 (image author's own).

I have introduced you to *invisible Zürichs*, a project organized by my artist collective, zURBS, since the work of this collective informs, constitutes, and illustrates the main focus of this book: to scrutinize how socially engaged art may challenge conceptions on the production of urban space within the neoliberal city.

By critically scrutinizing the linkages between art, participation, urban space, and neoliberal urbanism, the book explores first how socially engaged art may influence the ways we live in and think about our cities; and second how socially engaged art constantly negotiates and reflects the subtle power relations that exist between artists and their collaborators in urban contexts. In pursuing these aims through a practice-led approach, the book provides a nuanced account of the social functions of art based on critical perspectives relating to issues of urban politics as well as politics of collaboration, participation, and representation. I here define practice-led research as an interpretative engagement with artistic production and experience (Nimkulrat 2007). Such research feeds on practice as well as the practitioner's reflections. As such, it refers to a conscious exploration of the knowledge involved in the making and execution of socially engaged art, not only of my own practice, but also that of others.

Reflecting on my own experiences with facilitating socially engaged art in a series of European cities, as well as examples of socially engaged art by other artists in the USA, Africa, and Asia, the book demonstrates the challenges faced when struggling to adhere to artistic aims and visions, while at the same time working within various neoliberal and institutional constraints and expectations. I define neoliberalism in line with Pinson and Journel (2016) as 'the set of intellectual streams, policy orientations and regulatory arrangements that strive to extend market mechanisms, relations, discipline and ethos to an ever-expanding spectrum of spheres of social activities' (p. 137). These streams and orientations are by no means coherent, but are characterized by fuzzy, contingent, and ever-mutating processes of regulatory change. In other words, neoliberalism is not a stable and coherent order, but should be seen as always-contradictory processes of constantly evolving restructuring (Pinson and Journel 2016).

In some ways, critiques of the neoliberal paradigm may seem a tad passé. A decade ago, Neil Smith (2008) wrote that neoliberalism 'is dominant but dead' (p. 155), calling for political work that will make neoliberalism 'cliché in practice as well as in concept' (p. 157). Nevertheless, Smith also points to the importance of not discarding neoliberalism as an entirely overworked concept in favour of ambivalence or resignation. While the world may not any longer be witnessing new hegemonic neoliberal ventures, the filling in of commercial, institutional, and geographical interstices of the overall project is certainly prevailing. Hence, in line with Bruff (2014), one may see neoliberalism as 'living dead' – 'intellectually discredited, but apparently immovable due to the "absence" of feasible alternatives' (p. 114). According to Bruff, we are currently witnessing the rise of 'authoritarian neoliberalism' in which non-market institutions, such as the state, play crucial roles in maintaining its ideology. In consequence, institutions that were formerly viewed as a form of social protection against socio-economic restructuring could now well be the means to establish such restructuring. Examples here include the move from treating unions as equal partners to increasingly using collective bargaining negotiations to discipline labour; the general shift from welfare to workfare; the decline of catch-all political parties that seek to attract voters from different points of view; and finally the ideologies and the resurgence of the far right and authoritarian forms of government. As Bruff (2014, 155) points out, these shifts have weakened the forums for progressive and more radical politics that focus on collective interests as mobilized through trade unions, welfare programmes, labour parties, and ultimately the state. This weakening of forums, where the pressures emanating from neoliberal socio-economic restructuring can be absorbed and ameliorated, has resulted in the emergence of more coercive neoliberal processes. These processes do not only land in cities or impact urban governance; cities provide the fundamental material bases for these processes as built environments have become crucial to capital accumulation.

However, cities also provide fundamental material bases for the contestation of these processes. As Brenner et al. (2011) point out, 'the "urban question" famously posed four decades ago by Lefebvre, Harvey and Castells remains as essential as ever' (p. 226). Urbanization is rapidly transforming global landscapes and cities are increasingly seen as agents for change (Hajer 2016). Hence, cities are constantly looking for new trajectories to find solutions for pressing issues of our times, such as the housing crisis, the climate crisis, migration, the future of work, health and care, and social unrest due to austerity measures. These are all issues that are particularly manifest in cities. In response, a variety of collaborative, participatory, and community-based artistic practices that fall under the umbrella term of 'socially engaged art' have come to be valued as a crucial creative means of empowerment in the midst of urban transformation. Socially engaged art points to a form of collaborative and participatory art practice in which the artwork is a project that is co-produced between artists and participants. Often, the aim is to generate change by intervening in social relations. Common for these practices is that they do not only engage with a specific group of people, but also with their social and cultural concerns (Cartiere 2016). Socially engaged art is thus often firmly embedded in the place in which it is sited, grappling with its social conditions, political contexts, and unique histories.

Within these contexts and conditions, the often-contradictory dynamics that configure public engagement in socially engaged art are put to the fore. While the relationship between art and politics is nothing new, the debate on the social and political role of art has recently been 'revamped' as artists and creative workers increasingly engage in the social and political space of cities (Serafini et al. 2018). Influential terms such as Nicolas Bourriaud's 'relational aesthetics' have come to define socially engaged art as characterized by the social relations produced through a dynamic process rather than by a finished aesthetic 'product'. This characterization has contributed to triggering fierce debates on the apparent tension between the aesthetic and the social. This tension marks one of the major challenges for artistic practices that marks an orientation towards the social: the contradiction between seeing artistic practice as a presumably free aesthetic space and the social and institutional reality of art with all of its implicit exclusions. These exclusions are constituted by social and material dimensions relating to, for example, neoliberal institutional constraints, aesthetic judgement and understanding, artistic autonomy and issues of instrumentalization, and ethical questions concerning how art 'should' operate.

Rather than succumbing into totalizing narratives about how art practices are inevitably instrumentalized as they become part of neoliberal structures, logics, and ambitions, this book emphasizes the need to think more carefully about the institutional, discursive, and political frames within which this work operates. While many debates on socially engaged art situate it within neoliberal flows, this book attempts to find ways in which socially engaged art may transcend these flows. As such, the book makes a significant contribution to debates on public engagement and participation, querying how artistic practice can engage publics in their urban environment in order to identify, and positively and creatively mediate, power imbalances attached to imaginary and material aspects of urban space. Central to this contribution is the book's practice-led approach, which is oriented around the work of my artist collective, zURBS.

Introducing zURBS

zURBS was formed in 2011 by Sabeth Tödtli, Sander van Parijs, and myself. The three of us met during an MA programme in urbanism that involved studying at different universities in Brussels, Vienna, Copenhagen, and Madrid. During our studies, we encountered various urban collectives, social movements, activists, festival organizers, and community groups such as Citymine(d) (Brussels), SOHO in Ottakring (Vienna), Superflex (Copenhagen), and finally the 15 M movement, which occupied Puerta del Sol in May 2011 as we arrived in Madrid for our final semester. These initiatives worked with different forms of artistic urban interventions in order to open up the city as an inspiration for people to create alternative urban futures. Instead of developing finished works of art in the city, the city is here turned into a laboratory for imaginative experimentation (Loftus 2009). Encountering these initiatives filled us with inspiration and a firm conviction that we also had to do something. In this spirit, we formed zURBS. As Sabeth and I were living in Zurich at the time, we decided that zURBS would begin its pioneering work in this city. Nina Lund Westerdahl, an artist and architect living and working in Zurich, soon joined the team as Sander took on a more remote role in our activities due to starting a PhD on local democracy in Ghent, Belgium.

We defined zURBS as a socially engaged artistic practice that would focus on dynamic processes that use strategies for participatory urban enquiry and action that is grounded in the arts. 'The arts' and 'the artistic' are here committed to what Herbert Marcuse (1978) defines as 'an emancipation of sensibility, imagination, and reason in all spheres of subjectivity and objectivity' (p. 9). As such, 'art' is understood as 'a site where "new multi-dimensional knowledge and identities are constantly in the process of being formed"' (Rogoff 2000, cited in Hawkins 2011, 465). Hence, aesthetic experience offers 'the ability [. . .] to transform our perceptions of difference and to open space for forms of knowledge that challenge cognitive, social, or political conventions' (Kester 2011, 11). The artistic framework of zURBS' work can here be seen as a method for exploring and fostering an urban imagination premised on the multiplicity and instability of spaces, materials, and practices. Within this framework, users are active as interpreters, appropriating the story of the city for themselves and making their own story out of it. Hence, zURBS is based on a belief in the empowering creativity of collective action and shared ideas. The outcomes are not known beforehand, and therefore the projects are open to all kinds of inputs, approaches, and perspectives. In this way, zURBS aims to investigate and work with the social and material space of the city, using a method that suits its complexity. The intention is to enable urban inhabitants to identify and negotiate power imbalances attached to imaginaries regarding space, self, and other, and thus foster a reimagination of urban space.

I define urban space in line with the well-known definition by Massey (2005), as a social and contingent construction that is always in the process of becoming. Furthermore, urban space is the product of interrelations and a multiplicity of encounters (Massey 2005; Wilson 2016). The happenstance, liveliness, and risk of these encounters are what make the urban a site in which the boundaries of the 'givens' and what is normal are constantly negotiated (Lefebvre et al. 1996; Merrifield 2013; Wilson 2016). This negotiation within and interaction with urban space does not rely on its material and physical form alone, but also on a social and imaginative interchange that is open-ended and exploratory. This brings to the fore the role of collective imaginaries as part of urban processes and change. Important here is critical attention to how these imaginaries get materialized in urban form, and in turn how our urban environment informs collective urban imaginaries.

Urban imaginaries and technocratic visions

Urban imaginaries refer to the somatic and cognitive image we carry within us of the places where we live our everyday lives. It is what Andreas Huyssen (2008) defines as 'an embodied material fact' that is part of any city's reality: 'what we think about a city and how we perceive it informs the ways we act in it' (p. 3). Urban imaginaries are always and inevitably social, involving a multitude of perspectives and subject-positions differentiated by class and race, gender and age, and education and religion. These social relations that influence our urban imaginaries are materially and spatially constituted, pointing to the reciprocal relation between the material and the social. When talking about material environments and materiality, I here refer to a spatial understanding of cultural-materialism, for example in terms of how our relations, behaviour, and practices in urban space are limited or supported by power relations that are manifested in material forms and designs. Accordingly, the material environment plays an active role in directing our actions within it.

The relation between urban imaginaries and material environments seems particularly relevant today as contemporary society is increasingly acknowledged as a processuality, which is always full of material objects in complex relations and networks. Advertisements and social media construct and communicate collective imaginaries informing emotions and beliefs about the relation between the good life and urban form, based on, among other things, the high levels of performance from our cars, environments, infrastructure, and so on. More often than not, the terminology used describes products as if they were living beings actively taking part in our lives. At the same time, the development of data-driven and user-led systems (such as the smart city) claims to enhance our ability to participate in, and engage with, our material environment. However, the technology underpinning these initiatives is often portrayed as being objective and non-ideological, grounded in either science or common sense. This presents an image of user-led systems as neutral and politically benign, ignoring the social and political interests, speculations, and calculations that produce them (Kitchin et al. 2015).

This form of 'depoliticization' (e.g. see Swyngedouw and Wilson 2014) relies on a technocratic way of thinking about the city that has been critiqued by urban thinkers such as Henri Lefebvre. According to Lefebvre, technocratic solutions to urban problems are not in themselves liberating. Without addressing social relations and their transformation, these solutions remain superficial (Lefebvre 2003, cited in Pinder 2015, 38). Today, for example, waste is seen as an object of municipal management. As a manageable object that is somewhat external to society, waste is open to technical and institutional solutions. However, as Sarah Moore (2012) points out, by rejecting the argument that waste 'just is', one can deconstruct the category of waste and recognize the often unjust and highly exclusionary socio-spatial orders produced by waste in relation to, for example, the modernist shibboleths of cleanliness, hygiene, and sanitation.

It is a well-rehearsed critique of technocratic visions that they tend to portray the city as an organization of 'fixed' objects that are detached from social practices. As the urban critic Raymond Ledrut emphasizes (cited in Deutsche 1996, 52), the moment the city is severed from its social production and seen as an objective and physical entity, it appears to exercise control over the very people who control and use it. By insisting that an object 'just is', technocratic visions resort to arguments outside argumentation – a city is a city – and so decree in advance which approaches to the material environment are legitimate. Accordingly, through multiple legal, physical, and symbolic means, users of urban space attach certain 'schemes of perception' to spatial structures, determining what kind of use is good, just, and appropriate (Cresswell 1996). The urban environment here becomes an active force in defining appropriate practice, and thus shaping the processes of meaning-making that determine the social and material function of objects and buildings: a bench is for sitting, not sleeping on; a train station is for people on the move, not for people staying to seek shelter; and so on. Within these schemes of perception, the formation of alternatives to present conditions tends to be based on established classifications that propose causal or other relationships between given social attributes.

Challenging the 'givens'

This book argues that socially engaged art has the potential to call into question established classifications that are based on the apparent naturalness of dominant values and definitions, and in this way challenge what is presented to us as given. Socially engaged art can be seen as offering a way to step back from the everyday without losing sight of it, and hence imagining alternatives that are rooted in, but not limited to, the present. The critical potential of socially engaged art is here highlighted. First, in terms of how this practice may expand our view on cities by explicitly linking our material environment with the production of urban imaginaries. And second, in terms of how socially engaged art may advance an understanding of the urban environment as something that is constantly invented and performed through social and cultural practices. The book thus challenges neoliberal and technocratic visions that tend to portray the city as an organization of fixed objects that are detached from their social contexts. It argues that the moment the city is severed from its social production and seen as an objective and physical entity, it appears to exercise control over the very people who control and use it. The socially engaged art that I discuss here foregrounds that this art form has the potential to make explicit how the city should rather be seen as an environment formed through various, often conflicting, social practices produced by diverse social groups. Accordingly, the book offers a close investigation of how socially engaged art may open up a range of novel and productive ways of thinking about and engaging with urban space.

However, if the significant value of socially engaged art for exploring the production of space in cities is to be realized, there is a need for a nuanced account that warns against reducing the relation between this art form and neoliberal urbanism to any simplistic principle. By scrutinizing a range of projects that zURBS has conducted in Zurich, Basel, Monthey, and London since October 2013, I provide a personal insight into the doubts, frustrations, and failures, but also the opportunities, that zURBS has faced when working within a context of the neoliberal city. The structure of the book follows zURBS' trajectory of trial and error chronologically as each new iteration of a project provided new possibilities but also new problems, from the optimism and enthusiasm in *invisible Zürichs* (October–December 2013) – the first project conducted as part of this research – to the last project, *St. Clement's Utopolis* (May–June 2015), which ended in despair and disillusionment.

The book's practice-led approach foregrounds that the boundaries between artistic autonomy, social responsibility, and instrumentalized art are not clear-cut, and that creating any form of binary between them is unproductive. Where theoretical debates risk getting stuck in the binary opposition between social commitment and artistic autonomy, the case studies developed in this book carefully scrutinize the types of relations that exist in socially engaged art in terms of participatory approaches, the politics of representation, and political potentials. By investigating the subtle negotiation of these relations, the book provides a powerful articulation of the link between the artistic process and the co-constitutive shaping of subjects, knowledges, and worlds. The value of this investigation is that it combines questions around the forms of participation and representation that are enabled through socially engaged art with the question of how this work can elucidate and intervene in the reciprocal relation between spatial and material conditions and social relations and practices. This approach problematizes the idea of socially engaged artistic practice as an inherently liberatory space while at the same time pointing to the potential for this art form to provide avenues where productive change can be brought about.

The book furthermore demonstrates the potential of socially engaged art to assemble (new) publics around urban issues. This is both to enchant publics through processes of creative making, and to insert 'lay' stories into the academic and expert practices of urban planning and design. The transformative potential of collaborative activity in terms of being made strange to oneself, as one's perspectives and practices are put into question by others, is here foregrounded. To question zURBS' practice in relation to other artistic practices becomes relevant in this regard. I contextualize zURBS' approach to collaboration in relation to artistic practices in which the public is not necessarily co-producing the artwork or that adopt different collaborative approaches. In Part II, I discuss artistic practices with different approaches to engaging publics in the urban environment, from favouring artistic autonomy over collaboration to demonstrating the strengths and weaknesses of a participatory approach. This discussion provides the terrain for reflecting on my own practice in relation to the subtle relations of power and representation that exist between artists and their non-artist collaborators.

In the remaining parts of the book, the reader is offered a detailed analysis of socially engaged art from what Jane Rendell (2010) calls 'a place between' theory and practice (p. 12). What is developed in the course of this analysis is not just new knowledge, but also a set of reflections that actively elucidate and problematize the practices and processes of knowledge production. Following Rendell's argument, practice is here not reduced to an application of or inspiration for theory. Rather, the relation between the two is fragmentary and partial. As Rendell (2010) points out, theories here throw trajectories, or suggest paths, 'into practice, rather than being used as "tools" for "application"' (p. 10). The book thus suggests a productive use of theory, which takes place in a terrain that questions our methodologies and terminologies. This way, I unsettle relations among theory, scholarship, practice, and application. Practice-led research plays a key role here as it initiates a flexible research process that experiments with new abilities to create knowledge and emphasizes creative discoveries. My interdisciplinary approach, which applies cultural and urban geography through an artistic and performative framework, hence has the potential to contribute to the development of urban knowledge and the creativity in which this form of research is both done and disseminated. My focus on artistic practices and urban imaginaries can here be seen as a message as well as a medium of enquiry. In the current reductive regime of public funding for public knowledge, and plans for narrow and 'matter-of-fact' approaches to academic knowledge and its public impact (Daniels 2011), it is worth affirming the importance of creative explorations that are concerned not only with what is, but also with what might be.

Methods and structure

My research on and with the above-mentioned projects was guided by methodological approaches particularly suited to three aspects of zURBS' practice: the co-creation of, the interaction within, and the reflection on our work. Participatory action research (PAR) helped me rethink the settings of public participation in terms of the co-creation of zURBS' practice by various participants and co-organizing institutions. An ethnographic approach focusing on participant observation allowed me to analyse the interactions between the co-producers of our practice, such as the relations between the zURBS team and the participants and the locations in which the workshops took place, as well as within zURBS, and between zURBS and the organizers and institutions involved in facilitating the workshops. Finally, semi-structured interviews and more informal post-workshop conversations provided access to reflections and layers of experience that were not necessarily expressed by the participants in the workshops as such.

To address this multilayered practice and experience of socially engaged art, the book explores three sets of questions concerning issues of meaning-making, participation, and representation. The first set regards the processes of meaning-making in relation to the urban environment: How can artistic practice make explicit the ways in which our cities are transformed and come into being in the course of ongoing everyday practices? To what extent are our material surroundings and their associated meanings played out as parts of social strategies? How can artistic practices enable us to understand and scrutinize the practical and mental conceptions of urban space, and hence help us give meaning to the very existence of the material world around us? What are the insights, conceptualizations, and theorizations that artistic practice can offer us in this regard?

The second set of questions focuses on the participatory process of critically engaging with urban space: How can the participatory process within socially engaged art enable participants to directly engage with the complex social and material structure of their everyday urban environment in order to recognize, question, and challenge the circumstances that regulate their everyday actions and behaviours? How can we facilitate a participatory process that is not mobilized with specific aims and outcomes in mind, but that enables the participants to articulate their experiences of the city on their own terms, and this way offer them an experience of agency and empowerment in relation to their surroundings? How can we approach participation as a situated process that is produced in the circumstances of a specific setting? To what extent can socially engaged artistic practice open up a participatory approach that creates the conditions under which a broad variety of residents can articulate their intimate narratives, personal attachments, and uses of urban space so that these emerge as publicly relevant?

Finally, the third set of questions concerns the representation of people and their interests: How may socially engaged art, as a framework for representation, provide a space in which a diverse set of identities and interests relating to the city can be created, expressed, and negotiated? What kinds of participation are enabled within the particular frames of representation provided by the socially engaged art? What subject-positions are available for the participants in these representative frames? How can socially engaged art avoid classifying its participants into predetermined categories in which identities are made, but rather enable a form of emancipative self-representation that produces its own categories of classification?

These questions are examined theoretically in the first part of the book and with respect to detailed examination of six relevant artists' work in the second. In the third and fourth parts, the previously laid out theoretical and conceptual concerns are carefully thought through in a practice-led account of my own socially engaged art with zURBS.

In Part I, I set out the critical, theoretical, and intellectual context for socially engaged art. By mapping out a range of interdisciplinary debates in cultural geography, performance studies, critical urban theory, and contemporary art theory, this part stages the key terms to be explored in the book – meaning-making, participation, and representation, and their interrelations. Drawing on contemporary art theory, I question the focus on subject-centred relations in socially engaged art, and argue that this focus risks neglecting the relation between social processes and the material manifestations that shape them. I employ insights from art-geography collaborations in order to point to how a renewed sensitivity towards material environments is present in artistic practices that are developed in ongoing dialogues with the field of geography. Theories from performance studies aid me in this and illustrate how geographical understandings of materiality can be expanded by making explicit how materiality is constantly invented through social practices. Finally, I attend to the critical potential of socially engaged art, in terms of how the practice can open up a rethinking of participation and politics.

Part II critically engages with the practice of international artists in order to set up a framework for the practice-led reflections in Parts III and IV. Questions of the processes of meaning-making and collaboration in relation to the urban environment are here examined through the work of Richard Wentworth, Joseph-Francis Sumégne, Jill Sigman, Group Material, Ai Weiwei, and Jeanne van Heeswijk. These artists are all working in the context of large cities undergoing rapid urban change, and thus illustrate how artistic practice may intervene in the social and democratic issues resulting from this. This part has two interrelated focus points relating to processes of urban engagement and collaboration. The first is to develop theories about how artistic practice can help us reflect on how we use and give meaning to the material world in our everyday lives: How can artistic practice make explicit the ways in which our material environment comes into being and is transformed within ongoing everyday practices? The second is to scrutinize different collaborative approaches that artistic practice applies in order to engage publics in this process of meaning-making in relation to the urban environment.

In Part III, I link the previous focus on meaning-making and collaboration with practical considerations concerning processes of participation and appropriation in relation to my own practice with zURBS. The notion of 'urban democracy', which examines the possibilities for and limits of city dwellers to appropriate urban space so that it meets their various everyday needs, is here examined through concrete examples from the zURBS project, *invisible Zürichs* (Zurich, 2013). This project vividly foregrounds questions relating to the participatory process of critically engaging with urban space and asks how socially engaged art can enable participants to directly engage with the complex structures of their urban environment. The challenges and potentials for socially engaged art to negotiate taken-for-granted assumptions of the 'proper' use and definition of neoliberal urban space are here foregrounded.

Picking up where Part III leaves off, Part IV addresses the challenges concerning the representation of people and their interests in socially engaged art. This implies moving the focus from how socially engaged art enables participants to appropriate urban space through participatory processes, to scrutinizing how the space produced in the projects constructs certain subject-positions for the participants. The analysis reflects upon/uses as a base for discussion three projects by zURBS, Basel (2014), Monthey (2014), and London (2015), that illustrate the challenges for socially engaged art to adhere to artistic aims of producing poetic and multilayered experiences for participants, while at the same time working within various neoliberal institutional constraints and expectations. The ability of socially engaged art to provide a so-called 'free' aesthetic field removed from neoliberal influences is here problematized.

The conclusion brings together the practical experiences and theoretical reflections developed throughout the book in a discussion of the challenges and limitations of socially engaged art in terms of working within the neoliberal city. Above and beyond merely a synthesis of the key ideas of the book, the conclusion focuses on drawing out the promise of socially engaged art, but also the pitfalls and critiques. This discussion will centre on the notion of contradiction as a constitutive feature of socially engaged art. While I point to the challenges this inherent contradiction poses, I also illustrate how working with, rather than against, contradiction may provide non-binary perspectives that offer a rethinking of socially engaged art and its ability to combine utopian aspirations with concrete action, despite having to negotiate its complicities with neoliberal urbanisms.

Bibliography

Brenner, N., Madden, D. J., and Wachsmuth, D. 2011. "Assemblage Urbanism and the Challenges of Critical Urban Theory." *City* 15 (2): 225–240.

Bruff, I. 2014. "The Rise of Authoritarian Neoliberalism." *Rethinking Marxism* 26 (1): 113–129.

Cartiere, C. 2016. "Through the Lens of Social Practice: Considerations on a Public Art History in Progress." In *The Everyday Practice of Public Art*, edited by C. Cartiere and M. Zebracki, 13–27. London: Routledge.

Cresswell, T. 1996. *In Place. Out of Place*. Minneapolis, MN: University of Minnesota Press.

Daniels, S. 2011. "Geographical Imagination." *Transactions of the Institute of British Geographers* 36 (2): 182–187.

Deutsche, R. 1996. *Evictions: Art and Spatial Politics*. Cambridge, MA: MIT Press.

Hajer, M. A. 2016. "Foreword." In *Experimental Cities*, edited by J. Evans, A. Karvonen, and R. Raven, xvii–xix. New York: Routledge.

Hawkins, H. 2011. "Dialogues and Doings: Sketching the Relationships between Geography and Art." *Geography Compass* 5 (7): 464–478.

Huyssen, A. 2008. "World Cultures, World Cities." In *Other Cities, Other Worlds*, edited by A. Huyssen, 1–26. Durham, NC: Duke University Press.

Kester, G. 2011. *The One and the Many: Contemporary Collaborative Art in a Global Context*. Durham, NC: Duke University Press.

Kitchin, R., Lauriault, T. P., and McArdle, G. 2015. "Knowing and Governing Cities through Urban Indicators, City Benchmarking and Real-Dime Dashboards." *Regional Studies, Regional Science* 2: 6–28.

Lefebvre, H. 2003. *The Urban Revolution*, edited and translated by R. Bononno. Minneapolis, MN: University of Minnesota Press.

Lefebvre, H., Kofman, E., and Lebas, E. 1996. *Writings on Cities*. Oxford: Blackwell.

Loftus, A. 2009. "Intervening in the Environment of the Everyday." *Geoforum* 40 (3): 326–334.

Marcuse, H. 1978. *The Aesthetic Dimension*. Toronto: Fitzhenry & Whiteside.

Massey, D. 2005. *For Space*. London: Sage.

Merrifield, A. 2013. *The Politics of Encounter: Urban Theory and Protest under Planetary Urbanisation*. Athens, GA: University of Georgia Press.

Moore, S. A. 2012. "Garbage Matters: Concepts in New Geographies of Waste." *Progress in Human Geography* 36 (6): 780–799.

Nimkulrat, N. 2007. "The Role of Documentation in Practice-Led Research." *Journal of Research Practice* 3 (1): 1–8.

Pinder, D. 2015. "Reconstituting the Possible: Lefebvre, Utopia and the Urban Question." *International Journal of Urban and Regional Research* 39 (1): 28–45.

Pinson, G. and Journel, C. M. 2016. "The Neoliberal City: Theory, Evidence, Debates." *Territory, Politics, Governance* 4 (2): 137–153.

Rendell, J. 2010. *Art and Architecture: A Place Between*, 3rd edition. London: I. B. Tauris.

Rogoff, I. 2000. *Terra Informa: Geography's Visual Culture*. London: Routledge.

Serafini, P., Cossu, A., and Holtaway, J. 2018. "Introduction." In *ArtWORK: Art, Labour and Activism*, edited by P. Serafini, A. Cossu, and J. Holtaway, 1–25. London: Rowman & Littlefield.

Smith, N. 2008. "Comment: Neo-Liberalism – Dominant but Dead." *Focaal* 2008 (51): 155–157.

Swyngedouw, E. and Wilson, J. 2014. *The Post-Political and Its Discontents: Spaces of Depoliticisation, Spectres of Radical Politics*. Edinburgh: Edinburgh University Press.

Wilson, H. F. 2016. "On Geography and Encounter: Bodies, Borders, and Difference." *Progress in Human Geography*, 41 (4): 451–471.

Part I

Problematizing socially engaged art

Challenging urban imaginaries

> For those who pass it without entering, the city is one thing; it is another for those who are trapped by it and never leave. There is the city where you arrive for the first time; and there is another city which you leave never to return. Each deserves a different name; perhaps I have already spoken of Irene under other names; perhaps I have spoken only of Irene.
>
> (Calvino 1979, 99–100)

Confident that imaginative literature could legitimately intervene in society, Italo Calvino positioned his book *Invisible Cities* (1979) to foster urban change on individual and social levels, and to advance a rethinking and revisioning of how cities are imagined, designed, and built (Modena 2011). *Invisible Cities* provides an important trajectory for this book as it illustrates how the production of urban imaginaries can be seen as interventions in social and material processes of city-making. In line with the human geographer David Harvey, Calvino emphasizes that if cities can be imagined and made, they can also be reimagined and remade. This claim calls for a critical reflection on how collective imaginations can be materialized in spatial forms. Hence, Calvino set out to investigate the nature of the city by thinking about urban forms in terms of meaning-making, or of the city as a signifying system.

Invisible Cities is framed as a conversation in the garden of the ageing emperor, Kublai Khan, to whom the merchant Marco Polo comes to tell about his travels. In short prose poems, Marco Polo describes 55 imaginary cities. Polo's projections of these cities interrogate the structure of a vast array of microsocieties, unveiling the reasons for their forms and for the behaviours of their residents. For example, the city of Zaira contains all its own pasts like the lines on a hand, directing readers to the relationship between architectural space and the events of history. Clarice, the tormented city that has several times decayed and then burgeoned again, underlines how buildings and materials are adapted to different uses throughout the course of history. The trading city of Euphemia points to how cities are places of exchange not only in goods, but in words, desires, and memories. Perinthia, which on paper was envisioned as a paradise city, warns us against formal, abstract, and immobile perfection that on the ground is transformed to an unnatural, deforming experience.

These cities provide a defamiliarization that thwarts visual and cognitive automatism by dislodging the city from the usual habits of perception. The novel suggests that no city can be grasped in its present or past totality by any single person. Urban imaginaries differ depending on a multitude of perspectives and subject-positions. Polo observes the cities from inside or outside, from afar or close up, from the plains, desert, mountains, or sea. The gaze that readers follow is decentralized; it encourages us to see the city from unusual perspectives by focusing on small and marginal things (window sills, manhole covers, antennae, skylights, and so on) rather than focusing on the city's monumental features. The readers are invited to engage with these multiple perspectives as a means to reflect on and reimagine the city outside of the novel and their relationships to it. Hence, the imaginary cities, like urban imaginaries in general, are more than representations or constructions of the world; 'they are vitally implicated in a material, sensuous process of "worlding"' (Gregory 2009, 282) in which the subject creates a whole world for itself. What is put to the fore here is the continual interaction between, first, the city as a multifaceted, dynamic system consisting in combinations of psychological elements such as desire, memory, subjectivity, and the passing of time; and, second, between the city's layout, architectural look, functions, and its social aspects such as culture and history. Hence, *Invisible Cities* illustrates how urban imaginaries depend on a reciprocal process between the observer and her environment; the environment suggests distinctions and relations, and the observer selects, organizes, and endows with meaning what she sees according to her own purposes and adaptability.

Invisible Cities challenges urban imaginaries that relate to taken-for-granted spatial orderings of the world and provides alternative urban imaginaries that place in bold relief the constructible and destructible nature of social habits and urban structures alike. The potential of artistic practice to question how the world is organized, and this way open the possibility for changing that same world, is recognized by a wide range of scholars. As the French philosopher Jacques Rancière (2004) argues, art can be seen to foster an autonomous experience that suspends the domination of the 'system'. This is because the undecidability of the experience implies challenging familiar categorizations, such as established views, assigned usage, and the spatially constructed order – what Rancière terms 'the distribution of the sensible'. The distribution of the sensible is here understood as that which 'revolves around what is seen and what can be said about it, around who has the ability to be seen and the talent to speak, around the properties of space and the possibilities of time' (Rancière 2004, 12). This 'distribution of the sensible' is strongly linked to the distribution of spaces:

> What are these places? How do they function? Why are they there? Who can occupy them? [. . .] It is always a matter of knowing who is qualified to say what a particular place is and what is done in it.
>
> (Rancière 2003, 201)

According to Rancière (2009), art has the capability to contribute to what he calls 'a new landscape of the possible' by challenging this so-called 'distribution of the sensible' (p. 103). There is an egalitarian promise in this capacity of art as it relies on everyone's 'own political subjectivisation in the act of interpretation and reception' (Roberts 2010, cited in Dawkins and Loftus 2013, 671). As Rancière argues, aesthetic experience is shared by all, regardless of their individual skills, dispositions, and education (Dikeç 2012). In this regard, art may facilitate a process of invention: of new collectives, new political objects, and new conditions of possibility (McFarlane 2011, 210).

In this part, I introduce a set of debates that constitute the theoretical framework of this book. Drawing on contemporary art theory and critical theory, I discuss: (1) how socially engaged art can make explicit the reciprocal relation between the production of urban imaginaries and the material environment; and (2) how this form of artistic practice can work back on our everyday spatial ordering principles to suggest other ways of thinking and doing. Central for this discussion is to problematize an approach to socially engaged art that predominantly focuses on the subject-centred relations produced within it. I argue that this focus risks neglecting the relation between social processes and their material manifestations, and thus fails to investigate the relation between urban imaginaries and the material environment.

In the first chapter, I address the potentials offered by geographical approaches for thinking about the making of places, spaces, and worlds in socially engaged art. I further point to how theories from performance studies may foreground an understanding of materiality as being constantly invented through social practices. Finally, I illustrate how mobilizing art practices in geographical investigations might enable self-conscious forms of place-based intervention, providing transformative insights in terms of personal engagements with the city. The critical potential of socially engaged art is here foregrounded. However, I argue that this critical potential may also be compromised through collaborative processes that focus on creating, rather than questioning, the coherence of urban space. These criticisms and potentials of artistic collaborations raise questions concerning the politics of participation and representation.

The second chapter discusses these questions. Issues concerning participatory approaches as well as how socially engaged art relates to particular notions of 'politics' will here be scrutinized. Finally, I return to Calvino and the hope for change that he is providing us with at the end of *Invisible Cities*.

Bibliography

Calvino, I. 1979. *Invisible Cities*, 2nd edition. London: Pan Books.

Dawkins, A. and Loftus, A. 2013. "The Senses as Direct Theoreticians in Practice." *Transactions of the Institute of British Geographers* 38: 665–677.

Dikeç, M. 2012. "Politics Is Sublime." *Environment and Planning D: Society and Space* 30 (1): 262–279.

Gregory, D. 2009. "Geographical Imaginary." In *The Dictionary of Human Geography*, edited by D. Greogry, R. Johnston, G. Pratt, M. J. Watts, and S. Whatmore, 282–283. Singapore: John Wiley & Sons.

McFarlane, C. 2011. "Assemblage and Critical Urbanism." *City* 15 (2): 204–224.

Modena, L. 2011. *Italo Calvino's Architecture of Lightness*. New York: Routledge.

Rancière, J. 2003. "Politics and Aesthetics an Interview." *Angelaki Journal of the Theoretical Humanities* 8 (2): 191–211.

———. 2004. *The Politics of Aesthetics*. (2013). London: Bloomsbury Academic.

———. 2009. *The Emancipated Spectator*. London: Verso.

Roberts, J. 2010. "Philosophy, Culture, Image: Rancière's 'Constructivism.'" *Philosophy of Photography* 1: 69–79.

1 Criticisms, reimaginings, collaborations

A materialist critique

The importance and potential of socially engaged art for producing and expanding our urban imaginaries chimes with Lefebvre's (2003) call for an approach to urban society as a 'virtual object'. To Lefebvre, 'the virtual' relates to a dimension that has not yet been realized but is a horizon towards which we can move. This virtual dimension of urban society is important in order to interrogate the interplay between the imagined and the material in a way that encompasses the conditions of both the known world and the horizon of possible worlds. Taking the sensuous, embodied creativeness of mundane everyday social practice seriously in order to highlight how bodies are implicated in issues of social difference and power relations, and the ways space might articulate these, is central in this regard. New critical urban theories such as assemblage urbanism, as well as more-than-human geographies and the exploration of the non-representational, have pointed in this direction by valuing alternative forms of knowledge and understanding such as affect, sensation, and intuition. Debates here relate to how materiality accommodates the affective, the habitual, the technological, the excessive, and the processual (Farìas and Bendner 2011; Latham and McCormack 2004; Lorimer 2013; McFarlane 2011; Thrift 2008; Whatmore 2006).

I certainly recognize the importance of these developments for encouraging greater attention to creative practice, to embodied ways of being, doing, and knowing, and to the interplay between the imagined and the material. McFarlane (2011), for example, links the assemblage idea with the virtual dimension of the urban and a critical 'imaginary' (p. 219) that, in line with Lefebvre's critical theory, focuses on the relationship between the actual and the possible. However, I am concerned that there is a certain risk here of cutting out half the equation by sidelining the material conditions that are part of a city to begin with, as well as those produced through the encounters taking place within it. For example, Brenner et al. (2011) argue that certain orientations of assemblage theory, i.e. those that aim to work as an alternative ontology for the city (e.g. see Farìas 2011), has a tendency of focusing solely on the agency of the materials themselves, and thus leaving out the structuration of urban processes by, for example, capital, states, or social movements. As such, this new urban theory leaves unaddressed important questions regarding the broader contexts within which various actants are situated and operate, and regarding how, when, and why particular critical alternatives may be pursued under specific historical-geographical conditions, and why some visions and imaginaries are actualized over and against others. Accordingly, Brenner et al. point out that McFarlane's understanding of the virtual dimension of the urban is significantly different from the dialectical neo-Marxist approach to the virtual as advocated by Lefebvre. In McFarlane's account of the virtual, potentiality is seen as an exteriority that lies *outside* of the present assemblage, rather than being understood as historically specific or immanent to the material relations that are scrutinized. According to Brenner et al. (2011), this approach is problematic in that it does not offer a sustained account of the 'context of context' (p. 234). The need to engage with the political-economic structures and institutions in which materials are embedded is here foregrounded. Accordingly, Raco (2018) calls for a greater attention to the conditions that underpin the articulation of urban imaginaries, and how these are shaped by specific geographical contexts.

The critical planning and neo-Marxist spatial theories (e.g. Harvey 2000; Lefebvre 2003; Massey 2005; Soja 2000) that form the conceptual framework of this book help me scrutinize these conditions. Hence, the book is oriented around a cultural-materialist approach that offers an understanding of materiality as the expression of political and social commitment and practice, in which the materializing of an abstract idea is its observable instantiation in society. I here combine an attention to affective, sensuous, and embodied processes with considerations of symbolic and representational forms of meaning, as well as the material forces and relations of production that inform these processes. This cultural-materialist approach emphasizes the relation between subjective experience and its socio-material conditions. From this perspective, there is no 'outside' from which change, critique, resistance, or opposition can be articulated. Change is only enabled in relation to present structures and systems rather than as an escape from them. The potential field of political action in socially engaged art is here understood as not solely about mobilizing sites, representing various perspectives, and portraying multiple worlds in order to ensure an inclusive practice, but also about recognizing the opportunities and limits within these sites for enabling or constraining the creation of such multiple worlds.

Although taking into account the context of context, I am aware that the neo-Marxist and cultural-materialist approach provides a particular framing that focuses exclusively on the human constituents of the urban, and thus excludes the agency or proto-agency of other-than-human materialities in the city. In the current era of 'planetary urbanization' (Brenner and Schmid 2014; Lefebvre 2003; Merrifield 2013), it is clear that urban phenomena are always linked to and dependent upon places and entities that were previously imagined as the city's 'other' (e.g. nature, animals, and other living and non-living entities) (Metzger 2016). The contemporary city is increasingly seen as a natural-cultural system, an ecology, foregrounding the entangled fates of humans and non-humans, culture and nature. Accordingly, urban theorists, such as Jonathan Metzger (2016), call for a broadening of the horizon in terms of what constitutes a democratic and political urban subject, and who or what should be taken into account in the consideration of possible urban futures. Metzger's point is important, particularly today, as we are facing an ecological crisis that is very much caused by urban lifestyles. The need to dramatically rethink urbanism and its governance is needed more than ever. However, in focusing on how socially engaged art may play a role in reorienting the city away from being an engine of capital accumulation towards what Purcell (2014) terms 'a constitutive element in the web of cooperative social relations among urban inhabitants' (p. 149), the agency of humans is inevitably the focus of my discussion. While I do agree with Metzger's call for a more inclusive approach to urbanism, I take responsibility for the exclusion of other-than-human entities in the framing of this book. This is not to say that I discharge these entities for being irrelevant or that I reduce them to an externality of urban life. On the contrary, I do acknowledge their importance and ask the reader to keep them in mind in relation to my discussions on politics, participation, and representation. However, it is beyond the scope of this book to address this aspect directly, and I can only encourage other researchers to pick up the thread.

The subject-centred focus of socially engaged art

The neglect of socio-material contexts within certain strands of new critical urban theories can be seen in parallel to approaches to socially engaged art that sideline the material conditions of the practice and focus solely on the subject-centred relations produced within it. In the following, I will critique this subject-centred focus in relation to the historical development of socially engaged art.

The focus on subject-centred relations in socially engaged art can be seen in relation to art's negation of the artworks' object that became especially virulent in the late 1960s and 1970s. In their seminal text 'The Dematerialization of Art', John Chandler and Lucy Lippard (1968), point to a negotiation of the art object in favour of emphasizing 'the thinking process almost exclusively', resulting in 'the object becoming wholly obsolete' (p. 46). This development relates to the advent of Fluxus, performance and conceptual art in the 1960s and 1970s. Performance here exited the designated performance spaces and entered into, for example, urban contexts. Accordingly, artists shifted the attention from 'inside' the artwork to focus on the *meaning* of art as constituted through its context. The art historian and critic Rosalyn Deutsche (1996) terms this the 'site-specificity' of an artwork, alluding to 'an aesthetic strategy in which the context was incorporated into the work itself' (p. 61). In her book *One Place After Another: Site-Specific Art and Locational Identity*, Miwon Kwon (2004) points to an observed expansion of the site in locational and conceptual terms, meaning that cultural debates, theoretical concepts, social issues, political problems, institutional frameworks (not necessarily an art institution), communities or seasonal events, and so on may be deemed to function as sites. Hence, 'the site' becomes a function of the social. The relation between the artwork and its 'site', then, is not based on physical permanence, 'but rather on the recognition of its unfixed impermanence, to be experienced as an unrepeatable and fleeting situation' (Kwon 1997, 91). As such, the site of the artwork is increasingly dematerialized, as 'the operative definition of the site has been transformed from a physical location – grounded, fixed, actual – to a discursive vector – ungrounded, fluid, virtual' (Kwon 1997, 95).

Rendell (2010) points to how the land artist Robert Smithson's dialectic of 'site' (non-gallery) and 'non-site' (gallery) developed in the 1960s and early 1970s can be seen as one of the first explorations of relational sites through art practice. Smithson's definition of 'site' emphasizes its ungrounded and fluid qualities; the site is seen to have 'open limits' and 'indeterminate certainty' in contrast to the non-site, which has closed limits and 'contained information' (Smithson 1972, cited in Rendell 2010, 25). This definition implies that the site is favoured in the dialectical relationship between the two. Site-specific works, then, could be seen as a critique of the capitalist and exclusive gallery system by proposing other, less elitist and more inclusive (for example, public), sites for art. However, referring to Kwon, Rendell argues that site-specificity should not be seen as an automatic signifier of criticality. As Rendell and Kwon emphasize, even though the artwork is located outside the art institution, it is not necessarily free of institutional constraints and influences relating to, for example, funding and accessibility. Nevertheless, Kwon observes a lack of criticality in much site-specific work. She here points to an 'undifferentiated serialization' (Kwon 1997, 110), in which the different sites risk becoming generic by 'taking' one site after another without carefully examining the relationship between objects, people, and spaces that are produced through them. The emphasis on site as being ungrounded and dematerialized may here have the result that site-specific practices neglect the relation between social processes and their material manifestations, which, as Harvey (1973) points out, in turn shape these processes. As mentioned, a similar neglect can be found in discussions that focus on a subject-centred approach to socially engaged art.

One example in this regard is Bourriaud's contribution to conceptualize and institutionalize the socially engaged art of the 1990s with the notion of 'relational aesthetics', as mentioned in the Introduction. Bourriaud (2002) defines relational art as 'a set of practices which take as their theoretical and practical point of departure the whole of human relations and their social context, rather than an independent and private space' (p. 113). More generally, Bourriaud seems to make a separation between the social and the material, non-social world. Space is separated from the social by being seen as non-relational and individual/isolated, as opposed to the sphere of exchange and interactivity belonging to the social world. Space, then, seems to be locked in the position of a static container or backdrop that is to be *enacted* by the 'actors and extras' that make their way across it: 'It is no longer possible to regard the contemporary work as a space to be walked through [. . .] it is henceforth presented as a period of time to be lived through' (p. 15). And he continues: 'The exhibition may have turned into a set, but who comes to act in it? How do the actors and extras make their way across it, and in the midst of what kind of scenery?' (p. 74). In this context, the material environment risks being seen as fixed and static, while any kind of transformation only happens within what Bourriaud terms the 'materialism of encounter' (p. 18), meaning the bonds that link humankind together in social forms.

My problem with Bourriaud's *Relational Aesthetics* is not that he engages excessively with the social, but rather he does not engage sufficiently with the importance of the material. Rancière critiques Bourriaud for seeing relational aesthetics as 'arrangements of art that immediately present themselves as social relations' that are not mediatized by forms (cited in Bourriaud 2009, 1). In response, Bourriaud emphasizes that he does indeed take into account 'the concrete reality of the work' such as 'the colours, the disposition of elements in space, the dialogue with the exhibition space, the formal structure of the installation', and so on (p. 2). Bourriaud is here equating the material with reality, with 'giving an exact idea of what it is actually like' (p. 2). By creating this clear distinction between 'the material' as a realm of reassuringly tangible or graspable objects and the non-material social world of relational aesthetics, Bourriaud fails to take into consideration the virtual dimension of the material as advocated by Lefebvre. Accordingly, Bourriaud's approach risks constructing socially engaged art as a free and open space that unproblematically overcomes the demands of the social world instead of engaging with and negotiating material and social differentials (of power, resources, and authority).

As a result, socially engaged art may create what Kesby et al. (2007) call 'isolated islands of empowerment' that are not sustained within the very differently constituted spaces of people's everyday lives (p. 25). One quite literal example of this isolated approach can be found in Alex Hartley's project *Nowhereisland* (2012). The project consisted of a floating structure, resembling a part of the Arctic archipelago removed 'from the noise of the urban centres of the Western world' and conceptualized as 'a place at which citizenship can be conferred, ceremonial duties undertaken', and in which new and old stories would be gathered as the island travelled along the south-west coastline of England (Nowhereisland 2018). As Hewitt and Jordan (2016) observe, *Nowhereisland* was intended as a tool to galvanize people's imagination from outside of current political concerns, to think about the system and alternative futures from outside of that system. However, as viewers were encouraged to turn their back on space and time as they were imagining alternatives, the propositions suggested relied on singular, generic, and universal pictures of a supposedly stable 'good society'. For example, 'Every Nowherian has the right to be silent' and 'Every Nowherian has the right to be heard'. According to Hewitt and Jordan (2016), the abstract idea of *Nowhereisland* resulted in the fact that not much was at stake, and as a result the demands made were tame, not leading to any real exchange of opinion. Hence, *Nowhereisland* could be critiqued, in line with much utopian thinking, for being merely an idealistic exercise in daydreaming that seeks to escape the present (Pinder 2002). This idea of the good society as being somewhere 'out there' as a daydream that ignores the reality on the ground has nothing to do with the virtual dimension of the urban as advocated by Lefebvre. As Purcell (2014), citing Lefebvre, points out, this virtual dimension:

> is rather an extrapolation or amplification in thought of practices and ideas that are already taking place [. . .] practices and ideas that are inchoate, that have not yet come to full maturity, but are nevertheless being expressed, if only hesitantly, fleetingly, or inarticulately.
>
> (p. 23)

Hence, as Hawkins (2013) points out, to locate socially engaged art as a potential field of political action makes urgent a close investigation of these works and the dynamics of their entanglements with their concrete social and material sites. In order, then, for socially engaged art to open up a range of novel and productive ways of thinking about urban space, there is a need for an increased critical attention to the instabilities of material meanings and forms that allows for a subtler negotiation of the boundary between the 'real' and imagined. In the following, I will thus address the particular contributions offered by geographical approaches for understanding processes of meaning-making and for destabilizing places, spaces, and worlds. Theories from performance studies will aid me in this, as notions of performance provide useful insights into how the material conditions of urban space are produced and enacted.

Geographical and performative reimaginings of social space

The idea of space as a social construction gained ground in the 1970s, following the general rise of social theory, and Marxist approaches in particular (e.g. Lefebvre 1991), within various academic disciplines. Rather than being conceived as being outside human existence – as a backdrop against which human behaviour was played out – space was now seen as being socially produced and playing an active role in shaping social life. In the 1980s and 1990s, this approach was taken further by influential geographers such as Edward Soja, David Harvey, and Doreen Massey promoting an understanding of space not only as a social construction, but an understanding of the social and cultural sphere as also being spatially constructed. In other words, space was now seen to assume the role of a social actor influencing the way in which society works (Marchart 2012).

Peter Jackson's *Maps of Meaning* (1989) is an influential book that paved the way for new and richly interesting fields of enquiry in this regard. Jackson marks the emergence of a so-called new cultural geography that links the idea of the social with the cultural. This approach is influenced by a view of culture as insistently political and social, which had emerged from Raymond Williams, E. P. Thompson, John Berger, and Stuart Hall. This cultural-materialist orientation involves a shift in emphasis from culture itself to the domain of a form of cultural politics 'in which meanings are constructed and negotiated, and relations of dominance and subordination are defined and contested' (Jackson 1991, 200). Cultural texts and visual imagery are central to Jackson's book, extending research from the interpretation of finished texts to the process of their making. The focus on the construction of meaning is central to this book as it offers an understanding of materiality as the expression of political and social commitment and practice. The material environment, including its rules and representations, is here seen as embedded in the social and cultural contexts of which it is part. My cultural-materialist approach, then, focuses on cultural and social mediation in terms of meaning-making and representation. As Harvie (2013) points out, this approach emphasizes the embeddedness of art in its material and historical context as well as in broader social issues. In other words, art does not exist in a vacuum, but participates in the dissemination of ideologies, whether by being complicit in them or by challenging them.

My cultural-materialist orientation relates to a concern among geographers in exploring material culture (Jackson 2000) and broader material networks, relating structurally to significant kinds of social difference and injustice (Harvey 2000; Lees 2002; Philo 2000). Performance and performativity have increasingly been considered by geographers as important conceptual tools for a critical geography concerned to denaturalize taken-for-granted social practices that relate to difference and injustice (Gregson and Rose 2000). Performance can here be seen to constitute a methodological lens that enables scholars to analyse events as performance. Protests, citizenship, civic obedience, identity, and even the built environment are performed daily in the public sphere. To understand these as performance suggests that performance is 'a term simultaneously connoting a process, a praxis and episteme, a mode of transmission, an accomplishment and a means of intervening in the world' (Taylor 2007, 15). The term 'performativity' originates from linguistic studies, and the recognition that language does not merely describe things, but actually makes them exist. For example, by saying the phrase 'I declare war', the speaker enacts it. Performativity is often posited as providing an important connection between identity, power, and the construction of space (e.g. see Bell et al. 1994; Dewsbury 2000; Longhurst 2000; Thrift 1997). The body here becomes a performative site upon which multiple social identities are continually encoded and potentially resisted. The work of Judith Butler (1993) on the social construction of sexuality and gender has been particularly influential in this regard.

However, some performative analyses have a tendency to focus on 'the ways people can and do act with freedom to self-author, exercising agency, control and power through everyday acts of self-articulation and self-creation' (Harvie 2009, 45). Performativity is here premised on the idea of people as independent social agents with the freedom and ability to change their lives and society in general. This understanding risks seeing performance and performativity as offering an unmediated authentic relationship to the world, escaping the limits and demands of the social and material context. Nash (2000) points to how 'this idea of noncognitive embodied practice is somewhat at odds with the deeply social character of coded performances of identity in theories of performativity' (p. 657). For a critical approach concerned with the constructedness of the social and material world, then, it is vital to conceptualize social practices and material environments as in some sense produced by power, and not as removed from power's political script. As mentioned, new critical urban theories risk neglecting this aspect of power when focusing solely on the agency of materials themselves. I here share Nash's (2000) concern that:

> abstract accounts of body-practices and the return to phenomenological notions of 'being-in-the-world' arises also from the danger [. . .] that they constitute a retreat from feminism and the politics of the body in favour of the individualistic and universalizing sovereign subject.
>
> (p. 662)

This concern relates to Deutsche's critique of phenomenological readings of the city. Deutsche (1996, 63) argues that these readings place subjective experience outside the socio-material conditions of the city, and thus fail to take into account that the city is already ideological and contains more than just schemas of perceptive behaviour. Accordingly, I use the notions of performance and performativity to make explicit the processes in which materiality is constantly invented and comes into being within ongoing social and cultural practices and processes of meaning-making. The work of performance scholar Rebecca Schneider provides useful insights in this regard.

In her book *Performing Remains: Art and War in Times of Theatrical Re-Enactment*, Schneider (2011) unpacks the antagonistic relationship between performance and the archive that courses through Western historiography, on the one hand, and performance theory, on the other. The problem with seeing performance as an anti-archive, according to Schneider, is that the archive is reduced to a passive container of remains – as 'supposedly enduring materials' such as texts, documents, and buildings (e.g. see Taylor 2007, 19) – and therefore cannot assimilate the liveliness and contingency that is the constitutive trait of performance. As a result, the archive's role as a performative space for the live encounter with privileged remains is neglected, and we miss the opportunity to use performance as a means of interrogating and destabilizing the archive itself. As Jacques Derrida points out in *Archive Fever* (1998), the materials in the archive, or in our case the materiality of urban space, are given for their future (re)enactment ('what will have been'). Thus, the archive is a place for performative repetition, not stasis. It points to an instability and process where places and objects are predisposed to being in constant transformation. Mike Pearson (2010) illustrates this by discussing the manifestation of 'contemporary past' (p. 43) and privileged remains in urban space. These remains are marks and traces of our movements, actions, and encounters in the material surroundings: graffiti, footprints, handrails rubbed naked of paint, stains, dropped groceries, vomited kebab, and so on – attesting to our presence as well as our absence. The marks of physical contact with urban space are authentic traces of the performance of everyday life: 'the result of routine, tradition, habit, accident, event, social ritual, of long-term evolution and unconnected short-term ruptures and singularities, of nearness, of dwelling' (Pearson 2010, 43).

Seeing urban space as a place for performative repetition and live encounters with privileged remains hence enables a complex engagement with our surroundings. It forms the basis for the production of urban imaginaries premised on the invention, transformation, and instability of space, materials, and practices. Materiality is here seen as what theatre scholar Erika T. Lin (2012) points to as 'a contingent stability that is constructed through repetition and exists in comprehensible form only within a discursive nexus that gives it meaning' (p. 7). Lin uses an example from the theatre and its semiotic function, turning, for example, a chair into a throne in order to illustrate this potential. One may say that theatre's relationship to matter is secondary; the imaginative labour required to transform a chair into a throne is significantly less demanding than the physical and cognitive labour of the carpenter who carves a throne out of oak. However, by letting the semiotic and experiential processes through which meaning is produced come readily into view, theatre expands our view of materiality: it reveals the relation between practice and representation by elucidating how representations are inseparable from the broader social practices that authorize their existence.

As I have illustrated in this part, geography and performance offer theoretical approaches to how socially engaged art may expand our view of materiality and explicitly link our material environment with the production of urban imaginaries. These approaches emphasize the multiple practices through which cities are produced as a play of the actual and the possible and resonate with broader histories of critical theory and critical urbanism. According to Harvey, for example, we cannot pretend that we are not embedded and limited by the institutional worlds and built environments we have already created. However, the question of imagination cannot be evaded as – and here Harvey refers to an example by Marx – 'what distinguishes human labour and the worst of architects from the best of bees is that architects erect a structure in the imagination before realizing it in material form' (Harvey 2000, 153). Thus, imagination can be seen as the first step towards collectively producing our cities, but at the same time the future must be constructed 'not in some fantastic utopian mold, but through tangible transformations of the raw materials given to us in our present state' (Harvey 2000, 191). Therefore, as Harvey remarks, it is important that social and collaborative processes are seen as bound up with considerations of the constraints of spatiality and the qualities of space. As Kesby (2007) points out, geographical approaches may help us scrutinize how spaces of power and empowerment are entangled, and make us aware of the limits of participation and collaboration, as well as their possibilities.

Collaborative art and neoliberal urbanism – a complicated relationship

The complicated relationship between collaborative, socially engaged, and public art and the field of neoliberal urbanism is well rehearsed in literature. While art is valued for involving communities in development processes (e.g. see Van Herzele and van Woerkum 2008), this use of art is also criticized for integrating local values and expertise in urban projects as part of various 'creative city' initiatives to catalyse investment to cities and neighbourhoods (e.g. see Mclean 2016; Peck 2005). Deutsche (1988) observes how the new site-specific art that emerged as part of urban redevelopment strategies in the 1980s in New York was accordingly co-opted into technocratic and neoliberal visions aiming to use artistic practice as a tool to enhance an understanding of public space as an organic unity, thus denying the legitimacy of spatial contests. Accordingly, she argues, art became instrumentalized through collaboration with dominant forces, and thus lost its potential for political intervention. Instead of making the social organization and ideological operations of a space visible, this form of public art aimed 'to present as natural the conditions of the late-capitalist city into which it hopes to integrate us' (Deutsche 1996, 66). The focus on collaboration was here oriented around extending idealist conceptions of art to the surrounding city, rather than, for example, exploring the mechanisms by which power relations are perpetuated in the material environment. The consistent invocation of collaboration endowed art with 'an aura of social accountability' (Deutsche 1996, 67). Utility and social function thus became the principle yardstick for measuring the value of public art. In more contemporary debates, art critic Claire Bishop (2012, 12) points to how collaboration is seen as a sign of social responsibility due to its critical distance from the individualism of the neoliberal world order. Artistic practice that makes references to collaboration, participation, community, and collectivity is seen to work against dominant market imperatives by channelling art's symbolic capital into constructive social change rather than supplying the market with commodities. The urgency of this social task has, as Bishop points out, led to a focus on the accountability of art in relation to what it can do for society. Answers to this question include 'increasing employability, minimising crime, fostering aspiration – anything but artistic experimentation and research as values in and of themselves' (Bishop 2012, 13). Accordingly, many artists voice a lack of concern with private self-expression, and thereby express their opposition to the autonomy and privilege of art in favour of its social responsibility. However, while focusing on fostering social change by opposing neoliberal structures, this form of art may nevertheless be complicit with these structures. By addressing this paradox, this book provides a nuanced account that warns against reducing these kinds of collaborations to any simplistic principle.

From the practitioner's point of view, it is important to acknowledge that the boundaries between artistic autonomy, social responsibility, and instrumentalized art are not clear-cut, and that creating any form of binary between them is unproductive. There is, for example, a tendency for writers and critics to create a binary between art that is autonomous, and thus perceived as being inherently critical, and art that is instrumentalized, and thus perceived as being co-opted and inherently *un*critical. As the Dutch socially engaged artist Jeanne van Heeswijk (2012) puts it, 'Why do we have to talk again about this binary position when, in my opinion, autonomy and instrumentalization are no longer oppositional strategies?' (p. 78). According to van Heeswijk (2012), this binary presumes 'that working together with different partners such as local governments, councils, or social housing organizations invariably means that the artist is going to be instrumentalized' (p. 78). As a reply to this presumption she states, 'I like being an instrument that works on self-organization, collective ownership, and new forms of sociability. I like being an instrument that enables all of us to occupy the place in which we live' (p. 78). Nevertheless, more theoretical debates on socially engaged art risk getting stuck in the binary opposition between social commitment and artistic autonomy. This is particularly well illustrated in the polarized discussion regarding politicized collaborative art between art critics Grant Kester and Claire Bishop. Bishop (2006) argues that the perceived ethical responsibility of collaborative art compromises the artistic significance of these projects and their potentials for disruption, discomfort, shock, doubt, or sheer pleasure. For Bishop, disruption is particularly important in order to question the coherence of the city and expose social conflict. Hence, in her view, art should focus on antagonism, conflict, and disturbance as crucial elements of artistic experience.

Bishop's view is contested by Kester (2004), who emphasizes that the role of the artist in socially engaged work is to facilitate an ethical process where material and social conflicts and differentials (of power, resources, and authority) are sidelined in favour of consensual dialogue that depends on a given equality of voices. He accuses Bishop of advocating inward-looking elitist art that attempts to maintain cultural authority and artistic autonomy through its antagonistic approach. Accordingly, he argues that the role of the artist in socially engaged art is fraught with the risk of calling for democracy from a structure of social relations that enacts inequality. Bishop (2006) equates Kester's moral concerns with the demand by neoliberal governments for the arts to be socially inclusive (e.g. compulsory participation in a consumer society), privileging participatory art as a way to provide homeopathic solutions to problems that are systemic. She points to how the focus on social participation is viewed positively because it creates submissive citizens who respect authority and accept the 'risk' and responsibility of looking after themselves in the face of diminished public services. Thus, participation within this context will not change or even raise consciousness of the structural conditions of people's daily existence; it will only help people to accept these conditions. Equalling collaboration with social commitment, then, is, according to Bishop, problematic because it may lead to a situation where collaboration is celebrated as inherently 'revolutionary' and 'resisting', with little critical attention given to the meaning of this collaboration and its production. While Bishop and Kester accuse each other of placing in jeopardy the political power of art, they do agree on the importance of exploring the types of relations that exist in socially engaged art.

Conclusions

In this chapter, I have outlined a cultural-materialist approach that emphasizes the relation between subjective experience and its socio-material conditions. This approach recognizes that socially engaged art does not exist in a vacuum and that there is no 'outside' from which potentiality or resistance can be activated. Hence, in order to understand how socially engaged art may open up a range of novel and productive ways of thinking about urban space, it is central to scrutinize the relation between social processes and their material context. This calls for a greater attention to the conditions that underpin the articulation of urban imaginaries in socially engaged art, and how these are shaped by specific geographical contexts.

The combination of artistic practice and geographical investigations that I have outlined in this chapter points to how place-based interventions can be understood as ongoing interactions with material and social relations and provide transformative insights in terms of personal engagements with space and place. This combination combines an attention to affective, sensuous, and embodied processes with considerations of symbolic and representational forms of meaning, as well as the material forces and relations of production that inform these processes.

I have further argued that geography and performance offer valuable approaches to explore how materiality is constantly invented and performed through social and cultural practices, as well as to scrutinize the types of relations that are produced through socially engaged art. This exploration calls for a close investigation of the types of collaboration produced within this form of art. The socially engaged art practices that I discuss in this book offer such a close investigation in relation to issues of urban politics, as well as politics of collaboration, participation, and representation. This investigation raises an important question: how to attend to the critical potential of socially engaged art without uncritically celebrating the approach as 'revolutionary' or 'resisting', but also without succumbing to a totalizing narrative of instrumentalized participation that has entirely merged with the spectacle, and thus produces passive subjects with no agency or empowerment. In the next chapter, I will address this question in relation to notions of resistance, participation, and politics.

Bibliography

Bell, D., Binnie, J., Cream, J., and Valentine, G. 1994. "All Hyped Up and No Place to Go." *Gender, Place and Culture* 1 (1): 31–47.

Bishop, C. 2006. "The Social Turn: Collaboration and Its Discontent." *Artforum* 44 (6): 178–183.

———. 2012. *Artificial Hells*. London: Verso.

Bourriaud, N. 2002. *Relational Aesthetics*. Dijon: Les Presses du Réel.

———. 2009. "Precarious Constructions: Answer to Jacques Rancière on Art and Politics." *Open #17: A Precarious Existence. Vulnerability in the Public Domain*, 1–8.

Brenner, N., Madden, D. J., and Wachsmuth, D. 2011. "Assemblage Urbanism and the Challenges of Critical Urban Theory." *City* 15 (2): 225–240.

Brenner, N. and Schmid, C. 2014. "Planetary Urbanisation." In *Urban Constellations*, edited by M. Gandy, 10–13. Berlin: Jovis.

Butler, J. 1993. *Bodies That Matter*. London: Routledge.

Derrida, J. 1998. *Archive Fever: A Freudian Impression*. Chicago, IL: University of Chicago Press.

Deutsche, R. 1988. "Uneven Development: Public Art in New York City." *October* 47: 3–52.

———. 1996. *Evictions: Art and Spatial Politics*. Cambridge, MA: MIT Press.

Dewsbury, J.-D. 2000. "Performativity and the Event: Enacting a Philosophy of Difference." *Environment and Planning D: Society and Space* 18 (4): 473–496.

Farìas, I. 2011. "Introduction: Decentring the Object of Urban Studies." In *Urban Assemblages: How Actor-Network Theory Changes Urban Studies*, edited by I. Farìas and T. Bendner, 1–24. London: Routledge.

Farìas, I. and Bendner, T. 2011. *Urban Assemblages: How Actor-Network Theory Changes Urban Studies*. London: Routledge.

Gregson, N. and Rose, G. 2000. "Taking Butler Elsewhere: Performativities, Spatialities and Subjectivities." *Environment and Planning D: Society and Space* 18 (4): 433–452.
Harvey, D. 1973. *Social Justice and the City*. London: Arnold.
———. 2000. *Spaces of Hope*. Edinburgh: Edinburgh University Press.
Harvie, J. 2009. *Theatre and the City*. London: Palgrave Macmillan.
———. 2013. *Fair Play: Art, Performance and Neoliberalism*. London: Palgrave Macmillan.
Hawkins, H. 2013. "Geography and Art. An Expanding Field: Site, the Body and Practice." *Progress in Human Geography* 37 (1): 52–71.
Hewitt, A. and Jordan, M. 2016. "Politicizing Publics: A Social Framework for Public Artworks." In *The Everyday Practice of Public Art*, edited by C. Cartiere and M. Zebracki, 27–45. London: Routledge.
Jackson, P. 1989. *Maps of Meaning*. London: Routledge.
———. 1991. "The Cultural Politics of Masculinity: Towards a Social Geography." *Transactions of the Institute of British Geographers* 16 (2): 199–213.
———. 2000. "Rematerializing Social and Cultural Geography." *Social & Cultural Geography* 1 (1): 9–14.
Kesby, M. 2007. "Spatialising Participatory Approaches: The Contribution of Geography to a Mature Debate." *Environment and Planning A* 39 (12): 2813–2831.
Kesby, M., Kindon, S., and Pain, R. 2007. *Participatory Action Research Approaches and Methods: Connecting People, Participation and Place*. London: Routledge.
Kester, G. 2004. *Conversation Pieces*. Berkeley, CA: University of California Press.
Kwon, M. 1997. "One Place after Another: Notes on Site Specificity." *October* 80: 85–110.
———. 2004. *One Place after Another: Site-Specific Art and Locational Identity*. Cambridge, MA: MIT Press.
Latham, A. and McCormack, D. P. 2004. "Moving Cities: Rethinking the Materialities of Urban Geographies." *Progress in Human Geography* 28 (6): 701–724.
Lees, L. 2002. "Rematerializing Geography: The 'New' Urban Geography." *Progress in Human Geography* 26 (1): 101–112.
Lefebvre, H. 1991. *The Production of Space*, translated by D. Nicholson-Smith. Malden, MA: Blackwell.
———. 2003. *The Urban Revolution*, translated by R. Bononno. Minneapolis, MN: University of Minnesota Press.
Lin, E. T. 2012. *Shakespeare and the Materiality of Performance*. New York: Palgrave Macmillan.
Lippard, L. R. and Chandler, J. 1968. "The Dematerialization of Art." *Art International* 12 (2): 31–36.
Longhurst, R. 2000. "Geography and Gender: Masculinities, Male Identity and Men." *Progress in Human Geography* 24 (3): 439–444.
Lorimer, H. 2013. "Materialities." In *The Wiley-Blackwell Companion to Cultural Geography*, edited by N. C. Johnson, R. H. Schein, and J. Winders, 32–35. West Sussex: John Wiley & Sons.
Marchart, O. 2012. "Art, Space and the Public Sphere(s): Some Basic Observations on the Difficult Relation of Public Art, Urbanism and Political Theory." *Eipcp Webjournal*, 1–19.
Massey, D. 2005. *For Space*. London: Sage.
McFarlane, C. 2011. "Assemblage and Critical Urbanism." *City* 15 (2): 204–224.
Mclean, H. 2016. "Hos in the Garden: Staging and Resisting Neoliberal Creativity." *Environment and Planning D: Society and Space* 35 (1): 38–56.

Merrifield, A. 2013. *The Politics of Encounter: Urban Theory and Protest under Planetary Urbanisation*. Athens, GA: University of Georgia Press.

Metzger, J. 2016. "Cultivating Torment: The Cosmopolitics of More-Than-Human Urban Planning." *City* 20 (4): 581–601.

Nash, C. 2000. "Performativity in Practice: Some Recent Work in Cultural Geography." *Progress in Human Geography* 24 (4): 653–664.

Nowhereisland. 2018. "Welcome to the Embassy." Available at: http://nowhereisland.org/embassy/ [accessed 04.04.2018].

Pearson, M. 2010. *Site-Specific Performance*. London: Palgrave Macmillan.

Peck, J. 2005. "Struggling with the Creative Class." *International Journal of Urban and Regional Research* 29 (4): 740–770.

Philo, C. 2000. "More Words, More Worlds: Reflections on the Cultural Turn and Human Geography." In *Cultural Turns/Geographical Turns: Perspectives on Cultural Geography*, edited by I. Cook, D. Crouch, S. Naylor, and J. Ryan, 26–53. Harlow: Prentice Hall.

Pinder, D. 2002. "In Defence of Utopian Urbanism: Imagining Cities after the 'End of Utopia.'" *Geografiska Annaler: Series B, Human Geography* 84 (3–4): 229–241.

Purcell, M. 2014. "Possible Worlds: Henri Lefebvre and the Right to the City." *Journal of Urban Affairs* 36 (1): 141–154.

Raco, M. 2018. "Living with Diversity: Local Social Imaginaries and the Politics of Intersectionality in a Super-Diverse City." *Political Geography* 62: 149–159.

Rendell, J. 2010. *Art and Architecture: A Place Between*, 3rd edition. London: I. B. Tauris.

Schneider, R. 2011. *Performing Remains: Art and War in Times of Theatrical Reenactment*. New York: Routledge.

Smithson, R. 1972. "The Spiral Jetty." In *Robert Smithson: The Collected Writings*, edited by J. Flam, 270–296. Los Angeles, CA: University of California Press.

Soja, E. 2000. *Postmetropolis: Critical Studies of Cities and Regions*. Oxford: Wiley-Blackwell.

Taylor, D. 2007. *The Archive and the Repertoire*. Durham, NC: Duke University Press.

Thrift, N. 1997. "The Still Point: Resistance, Expressive Embodiment and Dance." In *Geographies of Resistance*, edited by S. Pile and M. Keith, 125–151. London: Routledge.

———. 2008. *Non-Representational Theory: Space, Politics, Affect*. London: Routledge.

van Heeswijk, J. 2012. "The Artist Will Have to Chose Whom to Serve." In *Social Housing – Housing the Social: Art, Property and Spatial Justice*, edited by A. Phillips and F. Erdemci, 78–89. Berlin: Sternberg Press/SKOR.

Van Herzele, A. and van Woerkum, C. J. M. 2008. "Local Knowledge in Visually Mediated Practice." *Journal of Planning Education and Research* 27 (4): 444–455.

Whatmore, S. 2006. "Materialist Returns: Practising Cultural Geography in and for a More-Than-Human World." *Cultural Geographies* 13 (4): 600–609.

2 Resistance, participation, politics

Resistance

As Pinder (2011) observes, there is a tendency for critics and writers to deploy celebratory and undifferentiated notions of resistance when discussing the potential for contemporary art practice to effect social change. According to Pinder, there is a need to think more carefully about the politics of this practice in terms of how it might be better understood as compromising and making do with powerful spaces rather than directly changing them.

The celebration of resistance among critics and writers can be seen in relation to a liberatory and oppositional political potential that geographers have long located in art. Art is here seen as a subversive force in terms of promoting an aesthetic engagement with space that is other than the normative apprehensions and uses of that space (Hawkins 2013). The perceived oppositional political potential of art has its roots in a long tradition of cultural theory treating the streets as the place for resistance, opposition, disruption, tactics, and protests, as, for example, in the works of Lefebvre, Georges Perec, and Michel de Certeau. It can also be seen in relation to the politics and poetics of the performative walking practices of the Situationist International.

The social uprisings of the 1960s and 1970s, in Central Europe and Paris in particular, provide an important framework for the focus on art's potential for resistance and opposition. In these uprisings, the street was seen as the characterization of the everyday, and thus framed in opposition to the central power, characterized by the enclosed offices of experts and ministers (e.g. see Castells 1983). De Certeau (1984) critiqued the top-down view of these experts and ministers. He argued that although it is a powerful tool for the planner to have a map or a helicopter view, this power is deadening as it leaves out the dynamics and energy of the city itself. From above, things look frozen: you are so much outside reality that it is hard to imagine what brings life to the city. In this regard, de Certeau discusses the idea of strategies versus tactics, emphasizing that space should be seen as a 'practiced place'. A strategy is here defined as relating to an already constructed, static, given place/structure, whereas tactics are the practices of daily life, which engage with and manipulate this structure. Massey (2005) critiques this conceptualization of strategies vs. tactics for introducing a dichotomy between power and resistance, and structure and agency, which divides the space of the city in two: the city structure versus the small tactics and resistances in the street. As Massey argues, this binary implies a conception of power as a static and monolithic order. Furthermore, it risks romanticizing a mobile 'resistance' of tactics and everyday practices as inherently political, thus neglecting a careful examination of the politics of these tactics and practices.

As this book illustrates, social struggles are far more complex and segmented than being inherently political or directed towards a monolithic power structure. The struggles of urban inhabitants include a range of different identities and political interests against an array of social and spatial structures (Purcell 2002). An important aspect of this complexity of social struggles and movements today is the rise of digital activism. As Castells (2012) points out, the Internet has created 'a new species of social movement' (p. 15). In the face of so-called 'Twitter and Facebook revolutions' (Schradie 2018), social platforms are seen as having the potential to enable networked clusters of people to coalesce, respond, and mobilize to amplify messages beyond individuals and specific communities (Kuo 2018). Circulating hashtags (e.g. #BlackLivesMatter, #ArtSoWhite, #SolidarityIsForWhiteWomen) is seen as helping make, for example, racial justice issues become more visible to publics who may not be aware or exposed to racial inequities. However, it is also argued that digital media has made citizens act less as members of an organization and more as individual users participating in activism (Bennett and Segerberg 2012; Earl et al. 2017; Schradie 2018). Social media is here seen to foster an extension of social engagement into the private sphere of individuals and their everyday, turning individual action into the performance and expression of the 'I', while partially losing the representative function of the 'we' (Milan 2015). Hence, while helping to organize to collective action, social media platforms may also work at the expense of fundamental group dynamics such as internal solidarity, commitment, and responsibility towards fellow citizens (Milan 2015).

This complexity of social struggles raises several concerns for artistic practices whose aim is to promote social change: How can artistic practices contest complex power systems without trying to reduce them to fixed structures that limit and constrain subjugated groups? How can artistic practices foster group dynamics in which the 'we' is more important than the 'I'? How can artistic practices avoid being merely reactive to what is presumed to be there already, and rather be generative of new identities and histories that are responsive to various productions of urban space? How can one avoid falling into the trap of too easily reifying conflict as *the* foundational characteristic of the political potential of artistic practice when trying to make visible what the dominant consensus tends to obliterate and obscure? How can one adopt a nuanced, rather than antagonistic, approach that addresses 'the varied abilities of artistic practices to challenge – or not – prevailing norms and power relations' (Pinder 2011, 688)?

In order to address these questions, it is important to be critical of what forms of participation and what kinds of urban imaginaries are produced in and through socially engaged art and to what effect. As mentioned in the Introduction, urban imaginaries involve a multitude of perspectives and subject-positions differentiated by class and race, gender and age, and education and religion. These differentiations point to the importance of a specific attention to 'intersectionality' (Crenshaw 1989), in terms of the relation between imaginaries and the power hierarchies that influences the ways in which individuals make sense of their lives (Raco 2018). While originally applied to the experiences of black women, 'intersectionality might be more useful as a way of mediating the tension between assertions of multiple identity and the ongoing necessity of group politics' (Crenshaw 1991; cited in Hopkins 2017, 2). As Raco (2018) points out, multiple perspectives and identities are held by subjects in a dialectical manner. In some contexts, socially engaged art may facilitate forms of transformative experience that results in a mutual understanding and a desire to reshape subjectivities. However, given the dialectical and fluid nature of identity formation, socially engaged art may also generate hostility and divergence with specific types of difference and with certain groups. Questions of participation are central to scrutinize these varied abilities of socially engaged art to produce mutual understanding, collectivity, and tolerance.

Scrutinizing participation and community

Participation has become something of a buzzword not only for artistic practices, but also in urban research and planning, and represents both a political potential as well as several challenges and pitfalls for socially engaged art. Whereas the call for participation is often dismissed as more rhetoric than substance, I am, in line with Maarten Hajer (2005), arguing that it is not so much participation itself that is the problem, but the very conditions under which the exchange of ideas has to take place. Thus, Hajer encourages us to rethink the settings of public participation: How can one create the conditions under which various groups are allowed to create situated and shared knowledge and understandings?

Participation is often regarded as a concept that is warmly persuasive and fulsomely positive, promising a world where everyone gets a chance to take part in making the decisions that affect their lives. Nevertheless, many scholars and critics, like Bishop, point to a bias in participatory practices. Even though participation is meant to, for example, facilitate policy deliberation and allow for the voices of the citizens and their associations to be heard, participation often tends towards practices that have strong exclusionary effects (see Hajer 2005; Mansbridge 1980; Young 2000). Critiques point to how participatory practices are often guided by norms of deliberation that 'implicitly value certain styles of expression as dispassionate, orderly, or articulate', thus excluding participants who do not conform to this norm (Young 2000, 6–7). Within planning theory, Judith Innes, one of the field's leading theorists, has argued that public participation antagonizes participants as they feel compelled to speak of the issues in polarizing terms to get their points across. Furthermore, she criticizes participatory processes for being nothing more than rituals to satisfy legal requirements (Innes and Booher 2000).

As an alternative to these exclusionary forms of participation, Leonie Sandercock (2003) outlines an 'epistemology of multiplicity', a planning imagination for the twenty-first century 'that is utopian and critical, creative and audacious' (p. 2) and that emphasizes the need to plan for multiple publics (p. 73). This form of epistemology acknowledges the many ways of knowing and doing that exist in addition to the instrumental, scientific, and technical ways of knowing that prevail in planning theory. Storytelling is here an important 'tool' as a form of knowledge production that enables people to appropriate the story of their city for themselves and make their own story out of it, as well as imagining themselves in different stories. In order for socially engaged art to create a setting that allows for these multiple shared knowledges and understandings to be expressed, it is necessary to rethink the social aspect of this practice.

A general critique of socially engaged art is that it often obscures the central issue at hand: the discursive construction of community itself. As Kwon (2004) points out, unquestioned presumptions tend to designate the community as a group of people identified with each other by a set of common concerns or backgrounds, who are collectively oppressed by the dominant culture and with whom socially engaged art seeks to establish a collaborative relationship. In this regard, Kwon expresses concerns about Kester's theories regarding socially engaged art. In Kwon's view, Kester seeks to essentialize cultural and social identity by seeing the objectives and identity of a coherent community as determined by its members before any encounter with outside individuals or groups, including community artists. This presumption is reflected in participatory approaches that aim at 'tapping into' some form of pre-existing local knowledge. Local knowledge here risks being instrumentalized as participatory tools are evaluated and utilized for their practical usefulness, effectiveness, or deliberative potential. Hence, participation is mobilized with quite specific aims and outcomes in mind, failing to see knowledge as a situated process that is produced in the circumstances of a specific situation (see Haraway 1988; Latour 1987; Shotter 1993; Thrift 1996).

The instrumentalization of participation is illustrated in the growing range of participation tools – among them those involving artists and artistic interventions – that have become available to help researchers, planners, governments, and communities to integrate local values and expertise in urban projects (Van Herzele and van Woerkum 2008). Within these participatory processes, local knowledge is often seen as a collective 'knowing from within' (Shotter 1993), developed in intimate familiarity and social interaction both with the situation at hand and with those who participate in that situation (Yanow 2004). As such, local knowledge is easily accessible to locals but considered difficult for outside experts to 'capture' (Van Herzele and van Woerkum 2008). Knowledge is here seen as a form of transmission whereby there is something on the one side, in one mind or body, that must be transferred into the other, to another mind or body.

The rise of civic tech, like, for example, user-led complaint management systems such as FixMyStreet, where people can contribute information about problems (e.g. potholes) in their streets, is but one example of this form of instrumentalization of participation. As Gordon (2016) points out, the values that are brought about by this form of participation implies a distrust in institutions and the idea of thc city as a functioning machine that you can tweak and hack. While designed to make citizens feel empowered by being able to operate outside of the immediate function of institutions, the citizen is merely the producer of data while the government is the expert technologist. The relationship that is established between citizens and governments and/or planners can be compared to what Rancière (2009) identifies as a spectator/actor relationship. Rancière critiques the negative connotations of spectatorship that may be implied by this relation, specifically regarding the spectator's role as either passive or active by virtue of the knowing performer.

From the politics of participation to participation as politics

Sandercock (2003) presents an alternative to the focus on how storytelling can function as a way for planners to 'tap into local knowledges' (p. 77). She here emphasizes the dialogue between planners and communities of interests. By paying attention to each other's stories, one may move forward to a situation of mutual learning through an equal empowerment that validates and respects each other's knowledge. This approach chimes with Rancière's call for an 'equality of knowledge' that blurs the opposition between knowledge and ignorance, activity and passivity, spectator and actor, by acknowledging that there are neither facts nor interpretations, only different ways of telling stories. According to Rancière (2007), this requires 'spectators who are active as interpreters, who render their own translation, who appropriate the story for themselves, and who ultimately make their own story out of it' (p. 280). However, in order to avoid falling into the trap of relativism, it is important to see these stories as partial and situated in the material, social, and political conditions in which they are produced. Relativism suggests that knowledge has no foundation other than personal and individual interpretations, and therefore 'true knowledge' is relative. Donna Haraway (1988) situates relativism as 'the perfect mirror twin of totalization' (p. 584), claiming that by promising a vision from everywhere (relativism) and nowhere (totalization) equally and fully, you deny responsibility and critical inquiry. Hence, Haraway advocates the concept of 'situated knowledges', which emphasizes embodiment and partiality as the conditions under which knowledge is acquired.

This situated approach to storytelling is an important constituent of Sandercock's 'epistemology of multiplicity'. Particularly important is the openness of this approach in terms of seeing knowledge as a process that is constantly produced, rather than as a resource that can be tapped into. As Eleanor Jupp (2007, 2837) points out, recognizing multiplicity is important in order to question and undermine the assumption that there are distinct realms of knowledge that exist prior to participatory research and/or processes. Thus, storytelling fosters a conception of knowledge as something performative, made intersubjectively within particular sets of social relations, times, and places (Jupp 2007; Pain 2004; Van Herzele and van Woerkum 2008). Accordingly, this approach points to an understanding of the social not as a final state that can be reached, but rather as a work in progress. As Deutsche (1996) states, 'Democracy and its corollary, public space, are brought into existence, then, when the idea that the social is founded on a substantial basis, a positivity, is abandoned' (p. 274). In line with Deutsche, Kwon (2004) advocates an art that reckons with the impossibility of community, and quotes the French philosopher Jean-Luc Nancy: 'There is no communion, there is no common being, but there is being in common' (p. 153).

Rancière (2007) takes the idea of the impossibility of community further in *The Emancipated Spectator*:

> In a theatre, or in front of a performance, just as in a museum, at a school, or on the street, there are only individuals, weaving their own way through the forests of words, acts, and things that stand in front of them or around them. (p. 278)

One therefore has to abandon the idea that, for example, socially engaged art is inherently communitarian – automatically making people part of a collective, in and of itself. Drawing on the political theories of Ernesto Laclau and Chantal Mouffe, Bishop (2004) conceptualizes 'relational antagonism', advocating that socially engaged art is being predicated 'not on social harmony, but on exposing that which is repressed in sustaining the semblance of this harmony' (p. 79). What Rancière, Deutsche, Kwon, and Bishop point to here is the difference between the conditions of the (im)possibility of society and the various attempts nevertheless to construct society partially.

In recognizing difference as an important part of being 'together', one may open up a form of participatory practice that is not based on the idea of a shared pre-existent condition or the formation of a common identity ('being us'). Rather than reassembling multiplicity in a common world, participation is here seen as the aggregation of singularities ('being together') by creating a multiplicity of worlds (Badiou 2007). These singularities should not be conflated with the self-fulfilling neoliberal subject that is led by the market to reproduce, expand, and reinforce competitive relations between themselves. Whereas these competitive relations foster *inequality*, the 'being together' that is advocated above is based on a drive to *equality* in which 'difference' is seen as a constitutive part of solidarity.

As Oliver Marchart (2011) points out, solidarity has for a long time referred to a mutual bond *within* a given social group or community – solidarity among those similar to each other. However, Marchart claims that this position is more about interest – or identity politics – and that the term 'solidarity' only makes sense where one declares oneself in solidarity with others who are not already part of the same community (e.g. solidarity by heterosexuals with gay marriage). In the new online ecology of hashtag solidarity (e.g. #MeToo, #BlackLivesMatter), this understanding of solidarity, as requiring establishing a relation with someone who precisely is *not* like me and does *not* share my own position, seems to be more important than ever. While this form of solidarity does have the potential to elevate and circulate discourse around important social issues, it also raises questions around solidarity for whom and for what.

In the face of terrorist attacks, Western citizens are quick to conjure 'imagined digital communities' (Stephens 2015) based on hashtags such as #JeSuisCharlie, which was formulated in response to the shootings at the satirical weekly newspaper *Charlie Hebdo* in Paris in 2015. This form of hashtag solidarity can be criticized for serving to solidify identity positions by inviting people to position themselves as either for or against Charlie. Accordingly, it reinforces a sense of 'heightened nationalism' (Butler 2006), which makes it impossible to oppose having to be 'with us'; you either share 'our' position, or you are 'against us'. Where some people easily feel comfortable with this response, it leaves others, especially minority communities, alert. In order to refuse these imagined communities of 'us' and 'them', Marchart's conceptualization of solidarity becomes particularly important. His understanding of solidarity requires that one establishes a relation of solidarity with someone who precisely does *not* share my own position. This requires an act of 'self-alienation' in terms of having to, at least partially, disidentify with one's own identity and position (Marchart 2011, 970).

Art may prove helpful in this regard as it 'is committed to that perception of the world which alienates individuals from their functional existence and performance in society' (Marcuse 1978, 9). This break with everyday habits and routines may have the result that, as Jenny Cameron and Katherine Gibson (2005) point out, 'new forms of subjectivity may emerge through unexpected shifts in the visceral and affective registers that free embodied practices from their usual sedimented patterns, creating opportunities to act on other possibilities for being' (p. 320). These shifts in identification may be promoted by socially engaged art, encouraged by the focus on imagination and creativity through positive affective and emotional registers. Accordingly, one is allowed to engage with the complexity of the world around us by recognizing difference, and at the same time reach out to one another in solidarity, acknowledging common beliefs and shared responsibilities. Solidarity can here be seen as forming the base of a reconstruction of political action that rests on a radical understanding of democracy as a way of living together in which people rule themselves in solidarity with each other (Marchart 2011). Democracy thus needs to take into consideration the contingent nature of all social and material relations.

The world is not based on a firm ground or ultimate principle, nor is it entirely without any ground or principle (we are not living in a void). Rather, it is based on what Butler (1992) calls 'contingent foundations'. These foundations are, as Marchart (2011) points out, plural and temporal, they can be reversed, and they have to be established against conflicting foundational attempts. A democratic regime, according to Marchart, is a regime that accepts, even promotes, the absence of an ultimate ground. That which is contingent opens up possibilities that things could have been otherwise. Democracy thus opens a politics, or political action, in which we seek to lay new grounds and do not doubt our ability to do so. However, the notion of democracy here stumbles into a contradiction: If we accept the groundlessness of democracy, don't we then undermine the basis of our political actions?

Whereas it is impossible to realize the democratic project once and for all, it is possible to supplement a final ground with 'contingent foundations' that are perpetually negotiated and (re)actualized. Democracy can in this regard be seen as a 'virtual object' – a horizon towards which we can move and a possible way of living together – as discussed in the previous chapter. Following this line of thought, Marchart (2011) argues that we have to 'develop a more flexible and yet determined notion of politics – one no longer afflicted by the question of scale, of intensity or the unconditionality of the act' (p. 971). The kind of politics that Marchart is criticizing here is first the so-called revolutionist 'grand politics' (e.g. Lenin seizing the revolutionary opportunity in the October Revolution), which is based on the 'phantasmatic idea' of a total break or rupture with a given situation. Second, he is sceptical of the so-called 'micro politics' that celebrate individual everyday activity as inherently political. According to Marchart, political acts are a (re)configuration of a specific space, a (re)framing of a particular sphere of experience or matter of concern, and a (re)thinking of what is taken for granted and demarcated as visible and invisible. Following this understanding of politics, all the cases that I discuss in this book are political acts. Common to these political acts is that they refuse what is presented to us as given, but at the same time they do not lose sight of what already is, and always begin from the activities in which people are engaged. This notion of politics is central to Lefebvre's method of transduction (cited in Purcell 2013, 21), which is a way to step back from the real in order to cut a path that leads beyond the actual world already realized and towards a possible world yet to come. This way of seeing political practice as a form of transduction is shared by Calvino (1979). At the very end of *Invisible Cities*, Kublai Khan realizes that all the cities that Polo has described are marked by a serious flaw: none of them are perfect, or even good:

> [Khan] said: 'It's all useless, if the last landing-place can only be the infernal city, and it is there that, in ever-narrowing circles, the current is drawing us.'
>
> And Polo said: 'The inferno of the living is not something that will be; if there is one, it is what is already here, the inferno where we live every day, that we form by being together. There are two ways to escape suffering it. The first is easy for many: accept the inferno and become such a part of it that you can no longer see it. The second is risky and demands constant vigilance and apprehension: seek and learn to recognize who and what, in the midst of the inferno, are not inferno, then make them endure, give them space.'
>
> (Calvino 1979, 127)

Conclusions

In this chapter, I have scrutinized the critical potential of socially engaged art in relation to notions of resistance, participation, and politics. I have argued that it is important not to treat socially engaged art as being inherently communitarian or political, but to carefully scrutinize the types of relations that exist in this practice in terms of participatory approaches and political potentials. As an alternative to exclusionary forms of participation, I have advocated an approach to participation that sees knowledge as a process that is constantly produced, rather than as a resource that can be tapped into. This participatory approach emphasizes solidarity as a way of recognizing and accepting difference, rather than trying to overcome it. I have pointed to how this understanding of solidarity may open up for political actions that strive towards democracy as a 'virtual object' that cuts a path leading beyond the world already realized and towards a possible and desirable world yet to come. This project of transduction is not an attempt to escape the present, but to seek out and nourish the wide range of minimal politics that are around us all the time. Building on the theoretical debates outlined in this part, in the following part of the book I will engage with the artistic practices of a range of international artists in order to push further the questions of participatory and collaborative approaches in relation to processes of meaning-making in and of the neoliberal city.

Bibliography

Badiou, A. 2007. *The Century*. London: Polity Press.

Bennett, W. L. and Segerberg, A. 2012. "The Logic of Connective Action." *Information, Communication and Society 15*: 793–768.

Bishop, C. 2004. "Antagonism and Relational Aesthetics." *October 110*: 51–79.

Butler, J. 1992. "Contingent Foundations: Feminism and the Question of 'Postmodernism.'" In *Feminists Theorize the Political*, edited by J. Butler and J. W. Scott, 3–21. New York: Routledge.

———. 2006. *Precarious Life: The Powers of Mourning and Violence*. London: Verso.

Calvino, I. 1979. *Invisible Cities*, 2nd edition. London: Pan Books.

Cameron, J. and Gibson, K. 2005. "Participatory Action Research in a Poststructuralist Vein." *Geoforum 36* (3): 315–331.

Castells, M. 1983. *The City and the Grassroots*. Berkeley, CA: University of California Press.

———. 2012. *Networks of Outrage and Hope: Social Movements in the Internet Age*. Cambridge: Polity Books.

Crenshaw, K. 1989. "Demarginalizing the Intersection of Race and Sex: A Black Feminist Critique of Antidiscrimination Doctrine, Feminist Theory and Antiracist Politics." *University of Chicago Legal Forum 140*: 139–167.

———. 1991. "Mapping the Margins: Intersectionality, Identity Politics, and Violence against Women of Color." *Stanford Law Review 43*: 1241–1299.

de Certeau, M. 1984. *The Practice of Everyday Life*. Berkeley, CA: University of California Press.

Deutsche, R. 1996. *Evictions: Art and Spatial Politics*. Cambridge, MA: MIT Press.

Earl, J., Copeland, L., and Bimber, B. 2017. "Routing around Organizations: Self-Directed Political Consumption." *Mobilization: An International Quarterly 22*: 131–153.

Gordon, E. 2016. "Meaningful Inefficiencies: Resisting the Logic of Technological Efficiency in the Design of Civic Systems." In *Civic Media: Technology, Design, Practice*, edited by E. Gordon and P. Mihailidis, 244–266. Cambridge, MA: MIT Press.

Hajer, M. A. 2005. "Setting the Stage: A Dramaturgy of Policy Deliberation." *Administration & Society 36* (6): 624–647.

Haraway, D. 1988. "Situated Knowledges." *Feminist Studies 14* (3): 575–599.

Hawkins, H. 2013. “Geography and Art. An Expanding Field: Site, the Body and Practice.” *Progress in Human Geography 37* (1): 52–71.

Hopkins, P. 2017. “Social Geography I: Intersectionality.” *Progress in Human Geography*, 0–11.

Innes, J. E. and Booher, D. E. 2000. *Public Participation in Planning: New Strategies for the 21st Century*. Berkeley, CA: University of California, Insitute of Urban and Regional Development.

Jupp, E. 2007. “Participation, Local Knowledge and Empowerment: Researching Public Space with Young People.” *Environment and Planning A 39* (12): 2832–2844.

Kuo, R. 2018. “Racial Justice Activist Hashtags: Counterpublics and Discourse Circulation.” *New Media and Society 20* (2): 495–514.

Kwon, M. 2004. *One Place after Another: Site-Specific Art and Locational Identity*. Cambridge, MA: MIT Press.

Latour, B. 1987. *Science in Action: How to Follow Scientists and Engineers through Society*. Cambridge, MA: Harvard University Press.

Mansbridge, J. J. 1980. *Beyond Adversary Democracy*. New York: Basic Books.

Marchart, O. 2011. “Democracy and Minimal Politics: The Political Difference and Its Consequences.” *South Atlantic Quarterly 110* (4): 965–973.

Marcuse, H. 1978. *The Aesthetic Dimension*. Toronto: Fitzhenry & Whiteside.

Massey, D. 2005. *For Space*. London: Sage.

Milan, S. 2015. “From Social Movements to Cloud Protesting: The Evolution of Collective Identity.” *Information Communication and Society 18* (8): 887–900.

Pain, R. 2004. “Social Geography: Participatory Research.” *Progress in Human Geography 28* (5): 652–663.

Pinder, D. 2011. “Errant Paths: The Poetics and Politics of Walking.” *Environment and Planning D: Society and Space 29* (4): 672–692.

Purcell, M. 2002. “Excavating Lefebvre: The Right to the City and Its Urban Politics of the Inhabitant.” *GeoJournal 58* (2/3): 99–108.

———. 2013. *The Down-Deep Delight of Democracy*. Chichester: John Wiley & Sons.

Raco, M. 2018. “Living with Diversity: Local Social Imaginaries and the Politics of Intersectionality in a Super-Diverse City.” *Political Geography 62*: 149–159.

Rancière, J. 2007. “The Emancipated Spectator.” *ArtForum* March: 271–280.

———. 2009. *The Emancipated Spectator*. London: Verso.

Sandercock, L. 2003. *Cosmopolis II: Mongrel Cities in the 21st Century*. London: Continuum.

Schradie, J. 2018. “Moral Monday Is More Than a Hashtag: The Strong Ties of Social Movement Emergence in the Digital Era.” *Social Media and Society 4* (1): 1–13.

Shotter, J. 1993. *Conversational Realities: Constructing Life through Language*. London: Sage.

Stephens, A. C. 2015. “The Affective Atmospheres of Nationalism.” *Cultural Geographies 23* (2): 181–198.

Thrift, N. 1996. *Spatial Formations*. Thousand Oaks, CA: Sage.

Van Herzele, A. and van Woerkum, C. M. J. 2008. “Local Knowledge in Visually Mediated Practice.” *Journal of Planning Education and Research 27* (4): 444–455.

Yanow, D. 2004. “Translating Local Knowledge of Organizational Peripheres.” *British Journal of Management 15*: 9–25.

Young, I. M. 2000. *Inclusion and Democracy*. Oxford: Oxford University Press.

Part II

Meaning-making

The reciprocal relation between the material environment and social practices

For public engagement

In the political climate of neoliberal capitalism, ideas from the world of business management have increasingly been applied to the organization of cities. The New Public Management approach of the 1990s focused on running the government like a business in order to shake up the old bureaucracy. 'Creative city' champions such as Richard Florida and Charles Landry underpinned assertions about the need for cities to be creative and innovative in order to attract international capital investment and compete on a global level. Today, 'the experimental city' has become the new sub-brand, aiming to differentiate and position cities in global knowledge hierarchies (see Karvonen et al. 2016). 'Living labs', 'innovation hubs', 'city deals', and 'green deals' have become the new lexicon of urban governance. These discourses of labs and experiments can be seen in relation to an influx of natural science expertise into urban matters. As Hajer (2016) points out, the urban as a domain of the social sciences is complemented by advances of a new 'urban science' that 'conceives of the city as an algorithm and tries to extend its claims using the newly available "big data"' (p. xix). Critics raise their concern with this new urban science in that there are quite a few remnants of positivist and technocratic visions in it (May and Perry 2016). While invoking experimentation may appear to challenge the status quo, it tends to replicate technocratic visions that understand the city as an organization of neutral physical objects rather than seeing it as a social form. These technocratic visions are nothing new in the urban discourse. As Deutsche (1996) observes, writing about New York a couple of decades ago:

> Urban public spaces are endowed with substantive sources of unity. Particular uses of space are deemed self-evident and uniformly beneficial because they are said to be based on some absolute foundation – eternal human needs, the organic configuration and evolution of cities, inevitable technological progress, natural social arrangements, or objective moral values.
>
> (p. 275)

Deutsche (1996) points to the appropriation of urban space as 'a strategy deployed by a distinctly undemocratic power that legitimates itself by giving social space a "proper", hence incontestable, meaning, thereby closing down public space' (p. 275). Public space is here turned into proprietary space that instigates the idea of a natural owner (a park belongs to families with children, and not to homeless people who disrupt the harmony). Accordingly, public space is seen as an organic unity, denying the legitimacy of spatial contests. Similarly, current talk of 'scientific rigour', 'resilience', 'upscaling', 'samples', and 'replicating' provides a relief from the ambivalence and complexity of cities through a selectivity that omits so much and so many from its process. The 'urban experiment' may take place in a context, but it relies on isolating crucial variables in order to replicate it across continents and time regardless of that context. As May and Perry (2016) point out, this raises a set of problems, as time, place, context, and character cannot be uncritically reproduced in the same manner in different settings. As a result, there is a process of disengagement with the complex and contested realities of urban life, assisted by representation through particular indicators, targets, and metrics. Examples include measures that emphasize waste as a manageable object that is separate from society through a focus on technical and institutional solutions that neglects its social implications, as mentioned in the Introduction.

Deutsche (1988) argues that the function of much artistic practice in the context of technocratic approaches is to reify as natural these instrumentally driven and econocentric conditions. Artists become part of urban development and design teams, producing works that are not just artworks, but also usable objects; street furniture such as benches, bike racks, and bollards designate places in the city for people to sit, stand, play, eat, and even dream. Through its usefulness, art is integrated into society as a public good. As I discussed in Chapter 1, with reference to Deutsche (1988) and Bishop (2012), this conflating of utility with social benefit is tricky. As Deutsche emphasizes, when function is limited to utilitarianism, social activity is constricted to narrow problem-solving so that the provision of 'useful' objects automatically collapses into a public good. Accordingly, artistic practice is expected to respond to urban questions by constructing images of coherent, harmonious, well-managed, and beautiful cities. The urban environment here becomes an active force in glossing over growing structural problems relating to inequality and accessibility. It is designed so as to reassure observers that the city might be free of divisions and problems, rendering urban inhabitants politically passive. How can socially engaged art enable us to recognize these forces and initiate a critical questioning and challenging of the material and political environment in the neoliberal city? This part of the book seeks to answer this question by drawing on my first set of questions concerning the processes of meaning-making in relation to the urban environment. I will argue that in order to scrutinize the ways that socially engaged art may challenge homogeneous and harmonic conceptions of urban space, there is a need for an investigation of people's engagements with the everyday materiality of the places where they dwell.

Ashley Dawkins and Alex Loftus (2013) point to how artistic practice has the potential to build what they call a 'relational sensuousness' (p. 665) as a form of engagement with the urban environment. Drawing on the interactive 'minor architecture' (p. 672) by the New York-based artistic collective Forays,

Dawkins and Loftus illustrate how processes of urban engagement construct sensory moments that move beyond utility and functionalism. Forays produces objects such as garden hammocks and playground swings out of discarded building materials. The objects are located in public space and demand a sensory engagement from the people who encounter them. By interacting with the objects, participants are immersed in the everyday rhythms of New York, but at the same time the works' 'out of place' positioning removes the participants from the routines of everyday life. Hence, the collective argues that participants' engagement with the objects produces a new sensory experience that helps them make sense of the world (cited in Dawkins and Lofuts 2013, 672). The questions posed by these sensuous experiences are 'not utilitarian – "what can this action do for our city?" – but processual – "how can we use what is all around us to encourage conditions of possibility?"' (Dawkins and Loftus 2013, 674). Similar processual questions are posed by the artistic practices that I will discuss in this chapter. These questions point to an understanding of urban space as both a resource and a terrain over which democratic struggles to create a new reality might be conducted.

Arts for urban change

Following the philosopher and cultural critic Walter Benjamin's argument that transforming the means of artistic production is central to artistic struggles for political change (Loftus 2009), I argue in this section that artistic work may open up the city as an inspiration for citizens to create alternative urban futures. Collaboration is central in this regard. As Dawkins and Loftus (2013) point out, the new sensory experience provided by artistic practices is closely linked to the idea of collaboration, in the sense that it holds the promise that everyone is capable of imagining these conditions of possibility. Collaboration is here not invoked in order to attest to the social function or responsibility of art, as critiqued in Chapter 1, but rather to imply a collaborative reclamation of the spaces that organize everyday life. Hence, processes of urban engagement are seen as ends in themselves, emphasizing a heightened awareness of one's surroundings and built environments that enables a deeper understanding of urban dynamics. Accordingly, collaboration challenges the uneven fragmentation, deskilling, and micromanaging of the workforce under the conditions of neoliberal capitalism, and opposes the technocratic vision of the city as a product of experts.

The chapters in this part have two interrelated focus points relating to processes of urban engagement and collaboration. The first is to develop theories about how artistic practice can help us reflect on how we use and give meaning to the material world in our everyday lives: How can artistic practice make explicit the ways in which our material environment comes into being and is transformed within ongoing everyday practices? The second is to scrutinize different collaborative approaches that artistic practice applies in order to engage publics in this process of meaning-making in relation to the urban environment. These focus points provide the terrain for reflecting on my own work in the remaining parts of the book, in relation to: (1) how artistic practice may work with the material world around us and our diverse experiences of it; and (2) the subtle relations of power and representation that exist between artists and their non-artist collaborators in terms of engaging publics in artistic practice.

While I draw on examples from a range of artistic practices, my focus is on the work of British sculptor Richard Wentworth, the Cameroonian painter and sculptor Joseph-Francis Sumégné, the American choreographer Jill Sigman, the former New York-based art collective Group Material, the Chinese artist-activist Ai Weiwei, and finally the Dutch visual artist Jeanne van Heeswijk. I have chosen to focus on these artists because their work elucidates different aspects of meaning-making and approaches to engaging publics in the urban environment, as well as in a collaborative process of art-making. The work of Wentworth, Sumégné, and Sigman is oriented around how artistic practice can help audiences reflect on the ways we use, give meaning to, and are influenced by urban space and its material surroundings. These practices centre on artistic autonomy over collaboration, and do less to elucidate the aspects of co-production between artists and non-artists. The work of Group Material, Ai, and van Heeswijk do more to elucidate participatory processes. Their practices centre on collaborative approaches that attempt to actively engage publics in engaging with their urban and/or political environment. Members of the audience are here turned into co-producers of the artistic practice, which is seen as a collaborative process in and of itself. The six cases complement each other in terms of outlining the different intents, constraints, and contexts that are crucial for understanding how we live in, experience, and interact with urban space.

The research that underpins this part draws mainly on primary accounts given by the artists in in-depth interviews with me and other researchers, and given in the artists' own documentation of their practice (e.g. in articles, exhibition texts, blogs, public commentaries). These primary accounts were supplemented by reflections on my own encounters with the artists' practices, as well as secondary accounts of the artistic process found in articles and reviews of the artists' work. Both primary and secondary accounts should be treated with caution as artists are not necessarily those who have the best understanding of what they are doing, and art critics and the media tend to portray artists and their work within certain interpretative contexts. However, the framing of these artistic practices within my own analysis and within the theoretical context of the relation between the material world and processes of meaning-making ensures the contextualization of these practices beyond existing interpretative frames.

Socially engaged art in a global context

By discussing artistic practices from across the globe, this part of the book seeks to extend the discussion on artistic practice beyond Anglo-American conceptions of neoliberal urbanism and the 'artscape' to include the geographies of Asia and the Global South. As Grant Kester emphasizes in the editorial notes of *FIELD: A Journal of Socially-Engaged Art Criticism*, founded by him in

2015, socially engaged art is a global phenomenon, and its criticism therefore needs to present work from all parts of the world. Ren and Luger (2017) adopt the notion of 'worlding' in order to challenge established horizons of critical artistic practice in and beyond particular cities and continents, such as the entrenched Euro-American focus of art history positioning New York and Paris as the heartlands of critical artistic practice. The approach of 'worlding', then, is a form of intervention in order to expand the world in and through which we look at art. In line with Ren and Luger (2017), my analysis of these international art practices does not aim to focus on the geographies of difference, but rather to focus on bringing the various artistic practices together and this way disrupt the assumption that, for example, Douala and New York are incommensurate sites of comparison. In other words, this is not an attempt at expanding the existing Western art history canon by simply introducing artistic practices from 'elsewhere' as if it is simply a question of geographic diversity. Rather, it is an attempt to partially dismantle this canon by bringing together artistic practices in a way that does not predetermine their positions.

The artists that I am looking at in this regard are mainly working in the context of large cities (and societies) undergoing rapid change, and thus illustrate how artistic practice may intervene in the social and democratic issues resulting from this. The material and historical environments range from New York in the 1980s (Group Material) to contemporary New York (Sigman); from London in the past three decades (Wentworth) to contemporary Brussels (van Heeswijk); and from Douala in the 1990s (Sumégné) to the contemporary digital sphere in China (Ai). I will not go into details about the ways in which the artistic practices are informed by these contexts, except for significant circumstances where these raise questions directly influencing how the artists negotiate their position, how they engage with urban and political space, and involve its residents in the process. To go into greater detail is beyond the scope of this book, and is something I will thoroughly discuss in the next part, where I focus more explicitly on the participatory processes of critically engaging with urban space in Zurich.

The cases will be discussed in relation to a range of ideas and concepts that are central to how we might understand and scrutinize the practical and mental conceptions of the urban environment. Building on the cultural-materialist approach I outlined in Part I, I use the cases to provide a detailed analysis of how the city is formed through various social practices. The ideas and concepts that I discuss in this regard include Lefebvre's notion of a socially lived space as a useful approach that enables the concretization of mental experience by focusing on the everyday and human practice within it. I also consider psychologist James Gibson's concept of 'affordances', which elucidates the potentialities of the physical materiality of urban space for particular sets of actions, and how these derive from essentially social practices. Finally, I discuss the influence of the surrealist movement starting in the 1920s, pointing to a long artistic interest in the found object, in rearticulating what is familiar to us, and in this way make everyday practices and experiences available for questioning and criticism. Drawing on these theories, this part points to how artistic practice may challenge the idea that there is one specific and pregiven representation of the world. As Ben Highmore (2002) argues, the search for the perfect fit between a form of representation and its object needs to be called off; no form of discourse is ever going to be appropriate to everyday life in its fluid complexity. Instead, the propriety of discourses needs to be refused in order to fabricate an alternative aesthetic for attending to the experience of modern life. Following Highmore's call, my case studies make explicit an urban imagination premised on the instability of space, materials, and practices, and explore the political and critical potential of object 'reuse'. Starting with the works and practices of Wentworth, Sumégné, and Sigman in the next chapter, I point to how artistic practice may foreground the critical practices that already exist in urban space, as well as creating interventionist actions that promote new ways of being in the city.

Bibliography

Bishop, C. 2012. *Artificial Hells*. London: Verso.

Dawkins, A. and Loftus, A. 2013. "The Senses as Direct Theoreticians in Practice." *Transactions of the Institute of British Geographers* 38: 665–677.

Deutsche, R. 1988. "Uneven Development: Public Art in New York City." *October* 47: 3–52.

———. 1996. *Evictions: Art and Spatial Politics*. Cambridge, MA: MIT Press.

Hajer, M. A. 2016. "Foreword." In *Experimental Cities*, edited by J. Evans, A. Karvonen, and R. Raven, xvii–xix. New York: Routledge.

Highmore, B. 2002. *Everyday Life and Cultural Theory*. London: Routledge.

Karvonen, A., Evans, J., and Raven, R. 2016. *The Experimental City*. London: Routledge.

Loftus, A. 2009. "Intervening in the Environment of the Everyday." *Geoforum* 40 (3): 326–334.

May, T. and Perry, B. 2016. "Cities, Experiments and the Logic of the Knowledge Economy." In *The Experimental City*, edited by A. Karvonen, J. Evans, and R. Raven, 32–46. London: Routledge.

Ren, J. and Luger, J. 2017. *Art and the City: Worlding the Discussion through a Critical Artscape*. London: Routledge.

3 Promoting new ways of being in the city

'Seeing the world as a Wentworth'

Richard Wentworth has played a leading role in British sculpture since the 1970s. His artistic practice revolves around 'the semantics of the everyday world, taking readymade and frequently incongruous objects and arranging them in a fashion that forces us to recognise the drama inherent in that which we too easily dismiss as routine' (Eastham 2011). Wentworth's interest in the unusual or counter-intuitive behaviour of things crystallizes in his photographic semi-diaristic reportages of the urban landscape as consisting of ready-made works that merit the same attention as more traditional art objects. One of his most renowned photo-series in this regard is *Making Do and Getting By*, which can be seen as a 'portrait of place' (Hawkins 2010, 815). The series is comprised of thousands of photographs, taken in the course of over three decades, of Wentworth's home territory, Caledonian Road in London. The area is framed through the arrangements of objects that he encounters in his day-to-day journeying around the neighbourhood. Witnessing the regeneration of Caledonian Road following the massive development of the nearby King's Cross Central, Wentworth's photographs portray a spectrum of discarded things and their changing value within the economies of the city. As a reaction to what he observes as an approach to urban space led by the desire for new and perfect materiality, Wentworth's photographs are haunted by a sense of nostalgia for skills and objects that are in decline, such as wooden roller shutters, hot-water heaters, or manual weighing scales. These objects bear witness to a time of 'industrial competence' when Britain, according to Wentworth, had the competencies not only to make, but also to mend things (cited in Hawkins 2010, 812).

However, Wentworth's oeuvre is also one of change and progression. As Anna Dezeuze and Julia Kelly (2013) observe, Wentworth lends his agency as an artist to the anonymous person in the street by appropriating found arrangements and 'handiwork' that tell of spontaneous and everyday reflexes and habits that respond to given practical circumstances. This way, Wentworth treats the everyday as a world of representation that can be articulated, reflected upon, and communicated. As Hawkins (2010) points out, the objects are revalued not for what they mean, but what they do, or rather what they could do and what people could do with them. Hence, the political potential of Wentworth's work lies not so much in interventionist actions that promote new ways of being in the city, but rather in a particular attention to critical occupations of space already in existence.

Wentworth's artistic practice is inserted into practical frameworks, embedded in in-the-moment practices relying on habits and routine practices of the body. These everyday distracted and tactile forms of habits and routines relate to what Benjamin (1968) describes as a 'tactile appropriation':

> Buildings are appropriated in a twofold manner: by use and by perception – or rather, by touch and sight. Such appropriation cannot be understood in terms of the attentive concentration of a tourist before a famous building. On the tactile side there is no counterpart to contemplation on the optical side. Tactile appropriation is accomplished not so much by attention as by habit.
>
> (p. 240)

As Graeme Gilloch (2002) argues, distraction is here seen as a common condition of habit formation and learning. It should thus not be regarded as simply inattention, positing distraction as a negative term relating to a common (surrealist) critique of the modern subject's impoverished and abstract encounter with the world. For Benjamin, distraction is rather a mode of cognition through which the public can understand and transform its own experiences. One requires habits while being distracted, but one is also distracted when habits are disturbed or interrupted. Distraction, then, is closely connected to attending to what is apparently insignificant, marginalized, overlooked, or neglected. Hence, it holds much power within it: 'it is a way of looking that feels its way around the place it finds itself rather than fixing that space with a distanced look' (Latham 1999, 463). Wentworth's work reflects this way of looking as he compares himself to a dog that 'sniffs out what is essentially opaque and resistant' (Warner 1993, 13). This approach is concerned with disturbing and defamiliarizing any notion of wholeness and unity. It undermines totalizing perspectives that see reality as a given, natural, and familiar environment. The aim is not to see new things, but see things anew: 'to make the act of perception performative rather than merely constative' (Sheringham 2006, 82). What particularly attracts Wentworth's attention in this regard are arrangements and organizations of disparate parts and objects (interview by Joe Kerr in Borden et al. 2000). As he points out:

> I am probably more interested in the gap between [objects], than I am in them [. . .] the way that it leaves spaces between itself and the other thing. So, this becomes dynamic [. . .] things encounter each other, which is what cities are: it is just this huge endless encounter.
>
> (personal interview, 20.10.2014)

The surrealist idea of juxtaposition can be seen as an important influence here. By juxtaposing disparate elements (fur and porcelain, lobsters and telephones), surrealism used the technique of collage to defamiliarize and make vivid the everyday, so that it does not drop below a level of visibility, but rather attracts attention.

The found object was understood to generate a continuous play of the familiar and strange, and mediate between the competing realities of the mind/imagination and the external world, dialectically reconciling the two (Mileaf 2010, 94–95). The rearticulation of the everyday does not, in this context, imply an invention of new systems of representations and articulations, but rather a reconceptualization of what is familiar to us. What emerges here is an understanding of objects that extends beyond their signification, to understand them as all kinds of message bearers, material processes, technical tools and devices, bodily capacities, habits, and skills (Whatmore 1999). This redefinition and functioning of objects is neither solely human nor wholly conscious, as it relates to the 'affordances' of the objects. The concept of affordances is worth some further discussion in order to better understand how Wentworth's work adopts a form of relational stance that may link the social and material world by emphasizing the material conditions of human activity.

Affordances and intelligence involved in practice

Wentworth's meticulous documentation of 'how people place things' as 'any form of adjustment or accommodation' (Dezeuze 2013, 281) points to the practical and reactive nature of human perception and cognition in certain material circumstances:

> There is no time for reflection [. . .], only a need to fulfil. The immediate fix of repairing a car bonnet with a piece of carpet [. . .], like the urgency involved in sticking a foot to keep a door open, imply a simple (more or less urgent) necessity to find a practical solution to a pressing problem.
>
> (Dezeuze 2013, 281)

The objects are here extracted from their contexts in order to be used in specific situations, like a boot that is used to keep the door open, or a bottle cap used as an ashtray, or a row of coats hanging neatly from the hooks provided by a fence. This uncovering of the material properties of objects that contradict, reorient, or exceed their original usage can be seen as what Wentworth has termed 'intelligence involved in practice' (Dezeuze 2013, 294). Gibson's (1979) concept of 'affordances' is helpful to understand this form of intelligence better.

Gibson coined the term 'affordances' in order to describe the potential of an object for a particular set of actions. As Carl Knappett (2004) points out, the term 'counterpoise[s] [. . .] the dominant Cartesian-derived view that human action inevitably involves reflection and conscious evaluation' (p. 44). In other words, it signifies that the world is not only of our own mentalist making.

However, Gibson's concept of affordances has been criticized for focusing solely on how these affordances are directly perceived, without involving any kind of social mediation (e.g. see Costall 1995, 2012). The affordances of the object are not accessible from their physical form alone, but also derive from essentially social practices. A stranger to a mailbox cannot, for example, determine it as such

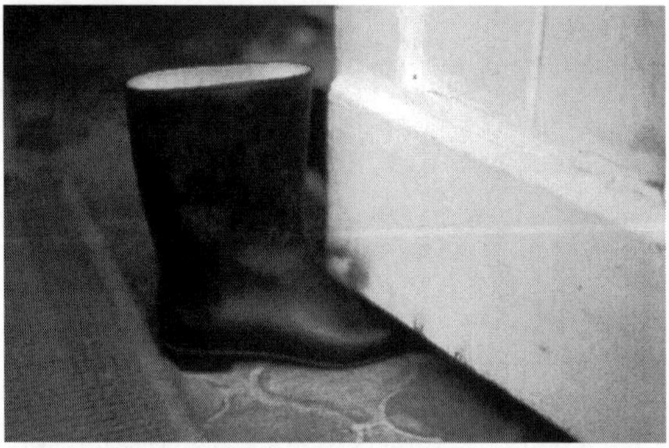

Figure 3.1 *Making do and getting by*, Richard Wentworth, undated (© Richard Wentworth; Courtesy Lisson Gallery; All Rights Reserved, DACS 2018).

merely by looking at it in sublime isolation from other people. We experience objects in relation to the community within which they have meaning. Hence, in order to adjust the concept, Alan Costall points to ways it should be socialized by highlighting the context of our behaviour: our material surroundings and their associated meanings are played out as parts of social strategies (Hodder and Hutson 2003). Objects are never neutral; they take on multiple meanings as they become part of various social practices in which we use a myriad of means to create new roles, to redefine existing ones, and to deny the existence of others.

The concept of socialized affordances elucidates the reciprocal relation between meaning-making and the material world around us. It sees space as produced and modified over time and through its use, invested with symbolism and meaning. Furthermore, it signifies that meaning can neither be 'privatized' mentally nor materially. Rather, meaning is based on the interdependence of body and world. Our potential for world-making, then, does not depend solely on our various perspectives fashioning portrayals of multiple worlds. It also depends on the opportunities and limits in our environments that enable or constrain the creation of such multiple worlds, in terms of, for example, collaboration, experience, knowledge, and culture.

In the research project *usus/usures* undertaken by the Belgian collective Rotor in 2010, the notion of 'wear' as a material phenomenon and agent that can influence behaviour is analysed in relation to these limits and opportunities within our environments. In an urban realm of new, perpetually replaced, pristine, smooth, polished, shiny, and glossy commodities, wear has often got negative connotations as a sign of ageing, breaking down, decline, deterioration, and vulnerability. The relationship with a neoliberal city sustained by new objects and the desire for perfect materiality is one of exclusivity and inhibition (Boniver et al. 2010). As the cultural geographer Tim Edensor (2007) points out, 'we are affectively and sensually alienated from the material world' (p. 226), and hence have problems apprehending the 'thingness' of objects, their material qualities, and their potentialities for manual apprehension. This alienation, according to Edensor, is caused by the regulatory measures in current planning, policing, and commodification of space. Large-scale, customized, Disneyfied, themed developments, shopping centres, festivals, heritage sites, spectacles, and gated communities constitute a host of designed realms that seem to produce familiar sensual experiences. Here, harsh, uncanny, and ambivalent sensations are kept at bay by the regulation of sensory intrusion and the production of moderated urban scenes. These developments have resulted in an entrenched form of urban habitus that has inured us to sensory overload as well as the vast complexity of everyday interactions.

This desensualization of space has social implications as it impedes people from associating themselves with the collective practice of preceding users showing confidence in the traces they leave behind; for example, the complicity in infringement when sticking a piece of chewing gum onto a spot littered with other pieces of gum. Or, in contexts where an action can be risky, wear or signs of acts that are repeated, and therefore in principle 'successful', will guide whoever subsequently encounters the same problem; for example the gradual forming of a path through a terrain that is difficult to traverse, or the double door where the one handle that is polished from use indicates that this is the side that will open (Boniver et al. 2010). The obsession with the new, removing all memory of other uses, challenges the potential of urban materiality to function as a medium for sharing and establishing relationships that brings one closer to the crowd. Rather than providing a form of 'security' in terms of making urban space legible, it singles one out, because the first alteration of an artefact carries with it responsibility; the artefact is moved from its intact, fixed, and immaculate state and into one of transformation. Take, for example, the decision to hang a shelf on the wall in a rented flat: if there are already holes in the wall, it might be easier to decide to hang the shelf there than if the wall has no holes in it. In this regard, wear can be seen as a messenger of content capable of influencing those who interpret.

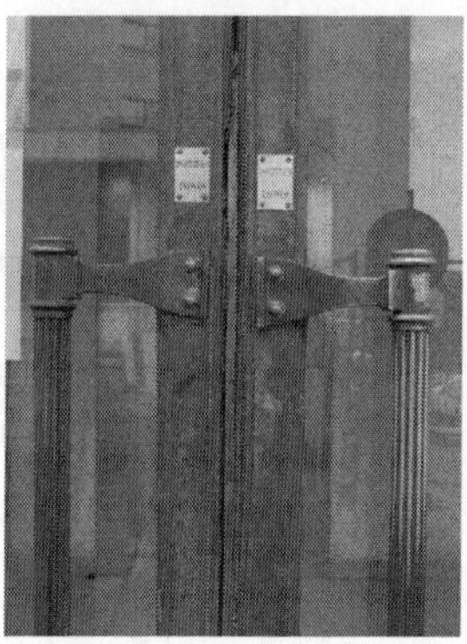

Figure 3.2 Door Handles, Rotor et al., 'usus/usures Etat des lieux – How things stand Brussels', Editions de la Communauté française Wallonie-Bruxelles, 2010.

The influence of wear

Wentworth points to the ways in which signs of wear influence us. As he explains:

> I am intrigued by all those practices which are actually world forming, and which in turn we respond to – how cars are parked affects how you are as a pedestrian – all those kinds of essentially urban conversations between people and objects.
>
> (interview with Joe Kerr in Borden et al. 2000, 389)

Wentworth is here attuning the eye to what Edensor (2007) calls an emergent aesthetics, 'one that cannot be fixed through endless maintenance but is constantly becoming different' (p. 223). However, this material fluidity of things does not mean that we can therefore do anything with anything (falling into the trap of relativism). As demonstrated in the research on 'functional fixity' (see Duncker 1945), the awareness of the social definition of an object, such as, for example, the shared definition of an object, may 'blind' us to its other relevant uses. Wentworth illustrates this in a conversation I had with him over a cup of coffee:

> It is like knowing that I could stir my coffee like that [takes my knife and points to the handle as a stirring-tool] and I have done that, but it is not necessary today because I have this tea spoon. [. . .] [People] wouldn't do that [use the knife to stir] because they are so regulated. [. . .] And then the coffee doesn't get stirred.
>
> (personal interview, 20.10.2014)

This observation regarding the *regulation* of our behaviour according to the awareness of the social definition of an object is highly relevant regarding current 'depoliticization strategies' that enforce the idea that there is one specific and pregiven representation of the world. Political issues are here often made to appear to be 'normal' or 'natural', and therefore not up for discussion or questioning (Wood and Flinders 2014, 161).

In order to elucidate the regulative social definitions of objects, Costall (2012) has modified the concept of affordances by distinguishing between 'affordances in general' and 'canonical affordances'. The term 'canonical affordances' points to 'the conventional, normative meanings of things, notably in relation to human artifacts' (Costall 2012, 90). For example, the objectivized and impersonal affordance of a chair: one *sits* on chairs. The social dimension of affordances here extends beyond the limits of objects and artefacts, pointing to how human activity is also socially and culturally transformed regarding how we conduct ourselves in specific contexts and situations. As Dezeuze (2013) observes, much of the pleasure in Wentworth's *Making Do and Getting By* derives from witnessing the deployment of objects and artefacts in ways that subvert these canonical affordances, such as, for example, a chocolate bar wedged in a wall alarm bell in order to silence it. In line with the surrealist technique of détournement, the expressions of the capitalist system are here turned against themselves: the mass-consumable unhealthy chocolate bar is effectively impeding the highly functional alarm bell to remind us that 'time is money'. What interests Wentworth with these forms of détournements is 'the "rub" between what he describes as a "formalism" – design, planning, control – and the "ripostes to formalism", the *bricoleur*'s appropriation and subversion of these given signs, codes, objects, or structures' (Dezeuze 2013, 295). Often these subversions are more ambiguous than the chocolate bar example. They range from convenient appropriations (using bricks to stand on to get a better view), polite messages (a single piece of tape discouraging us from opening the lid of a mailbox), and invitations (a sign inviting us to press a bell 'for a long time'), to more aggressive signals set up by arranging objects in certain ways (a chair and a wooden stick blocking a staircase).

These appropriations and subversions focus on potentiality and transformation, and foreground an aesthetic where the world is in the making (Hawkins 2010, 815). This focus is perhaps particularly explicit in cities of the Global South, such as Douala, where definitions of power are not always readily apparent, or are arbitrary, and urban survival relies on one's own ability and responsibility to keep things open and possible (Simone 2004). However, as I point out in the following discussion on the public artwork *La Nouvelle Liberté* in Douala, fetishizing chaos and openness is problematic in that it may render society unrepresentable, making it hard to orientate and/or seek out alternatives. In this context, artistic practice may play an important part in giving people a sense of direction and legitimization.

La Nouvelle Liberté, hodgepodge, and trash aesthetics

The construction and erection of *La Nouvelle Liberté*, a 12-metre-high sculpture assembled with bits and pieces of discarded objects found throughout the city, was a significant event in Douala, the economic capital of Cameroon, in the 1990s. The sculpture was made by the Cameroonian artist Joseph-Francis Sumégné, and initiated by *doul'art*, a centre for contemporary art established by Marilyn Douala-Bell and her husband Didier Schaub in 1991. This was during a height of violence and tension in Cameroon due to economic crisis combined with rising calls for democratization (Tambi 2015). In response to these tensions, the centre was formed to 'reaccustom people to having peaceful relationships' by inviting residents to experience artistic creation 'because this has always been a way in which we can share things' (interview with Douala-Bell, cited in Makongo 2018). Furthermore, the centre encouraged artists to leave their studios and 'give meaning to the city' (Makongo 2018). This latter aspect is particularly important in the context of Douala. As Simone (2004) points out, the regime that ruled Cameroon in the 1980s and 1990s decreasingly invested in the definitional aspects of rule, such as clearly defined jurisdiction, zones, policies, and sectors, and rather allowed unregulated practices of accumulation to unravel the structures on which social gravity once relied.

The practice of keeping things open and possible in situations where the definitions of power are not readily apparent or are arbitrary is often seen as a characteristic of African cities (Simone 2004). African urban life is accordingly often portrayed as having a love affair with 'hodgepodge', in terms of the incessant throbbing of the close proximity of hundreds of diversities in which provisional orders are hastily assembled and demolished. While this sense of throwing things together can be seen to give many cities their appearance of vitality, representing everyday space as a pell-mell of different worlds colliding is also problematic.

As Massey (2005) observes, 'it is popular today to revel in the glorious random mixity of it all. Chaos is taken to be a form of rebellion against over-rationalisation and the dominance of closed structures' (p. 111). This easy critique, which frequently associates order and uniformity with 'planning' or 'the state', tends to forget how the market or other non-state social forces hide their powers behind 'the new love affair with chaos' (Massey 2005, 112). Accordingly, Miranda Joseph (2002) observes how 'the association of homogeny with hegemony is a false one, as the explicit promotion of diversity in contemporary capitalism demonstrates' (p. 61). The artistic aim of breaking free from rational and homogenized society by dismantling the barriers that separate everyday life from art and re-enchanting the world through a focus on play, desire, creativity, and imagination has been rendered into new forms of capital accumulation through, for example, the cultural industries and strategies of city branding. As Steve Pile (2005) points out, cities have increasingly become part of the machinery of commodified dreams and desires: 'dream homes with dream mortgages, dream loans for dream holidays, dream hair to go with dream lingeries' (p. 167).

This focus on dreams, imaginations, chance, and the pell-mell of colliding worlds ensures the impossibility of closure and indeterminacy of space, ultimately rendering it unrepresentable (Massey 2005). Whereas this unrepresentability is praised for having a liberatory potential due to its 'freeing' diversity, it may in fact reproduce current structural conditions rather than changing them: by privileging an understanding of society as unrepresentable, celebrating what cannot be expressed rather than what can, capitalism challenges the possibilities for a sustained critique of the established order of things (Joseph 2002). Hence, present structural/social conditions make it hard to articulate or even imagine alternatives. This, however, does not mean that interventions that try to resist closed and homogeneous conceptions of space possess no value today; rather, it reinforces the importance of scrutinizing these particular engagements with the physical materiality of urban space in terms of how they address the relationships between objects, people, and spaces.

In the case of Douala, a context where it is often unclear who has the right and ability to do what, there is a pervasive anxiety on the part of urban residents in relation to who they can live and work with, who they can talk to, and what kind of collective future they can anticipate (Simone 2004). According to Bell-Douala (Makongo 2018), *La Nouvelle Liberté* played a central role in providing a sense of orientation in this regard. The statue was made out of scrap metals, car parts, and discarded tyres that Sumégné had found rummaging garbage cans and sites across the city. In Bell-Douala's view, Sumégné's focus on valourizing recycling and salvaging – a way of saving money that people were ashamed of and did not want to acknowledge as a part of the household economy – contributed to legitimizing these endeavours and even making people proud of what they do. The statue was furthermore placed in a traffic circle marking the most important node in the circulation of traffic across the city. The site was especially known for hosting 'illegal' commerce, such as sales of second-hand clothing by youths that persistently resisted being chased off the site by the police. In a collective statement after the statue was erected, the clothes sellers proclaimed, 'The work is like us . . . therefore we are always here . . . now we can go' (Simone 2004, 113).

In order to further understand the perceived emancipatory potential of *La Nouvelle Liberté*, it is helpful to look at Sumégné's practice in relation to what Highmore refers to as Walter Benjamin's 'trash aesthetics' (Benjamin 1999; Highmore 2002, 60). Trash aesthetics can be used radically and critically to attend to the everyday through its focus on the detritus and backsides of modernity. It may, for example, provide an alternative to the modern capitalist focus on progress and its celebration of the new. As Highmore (2002) points out, it was particularly the object as a ruin that fascinated Benjamin, as the object's ruination or 'after-life' facilitates a destruction of deceptive appearance. A contemporary example of trash aesthetics is the work of the Palestinian conceptual artist Bisan Abu-Eiseh. For his installation *Playing House* (Ramallah, 2008–2011), Abu-Eiseh collected objects from demolished houses in Jerusalem. The objects were displayed in vitrines with labels describing where the object came from, the date of the demolition, and the number of people displaced from the house. Abu-Eiseh explained that it was the process of giving these things value that particularly interested him (personal interview, 16.09.2014). Value is here not determined by the resale potential, provenance, functionality, or prestige of the object sustaining a disinterested or aesthetic contemplation (Bonnett 1992). Rather, the value of the object is its ability to tell stories about the people affected by the present developments in Jerusalem, in which over 8,000 Palestinian homes have been demolished since the 1960s, and their memories and experiences of loss in this regard. Christopher Woodward (2002, cited in DeSilvey 2006) observes how people, when confronted with ruins and remnants, generate creative resources through the sympathetic association between structural incompletion and imaginative invention. Degraded artefacts may in this regard contribute to alternative interpretive possibilities. Indeed, Simone (2004) observes how local civic groups repeatedly wanted to offer their interpretations of *La Nouvelle Liberté* and to know whether their interpretations made sense or were valid in terms of thinking about the sculpture.

The affirmative excitement of *La Nouvelle Liberté* was, however, by no means universal. One of the local newspapers, *El Limbi*, declared it 'A monument in shit', referring to the terrible conditions of the road and site of the roundabout on which the sculpture was situated. Instant repair works were set in motion, but the newspaper continued its criticism in a series of articles condemning the sculpture for not being aesthetically 'beautiful' and not being made with precious new materials as were public monuments in European cities (Simone 2004). Douala-Bell responded that the sculpture represented not *European* public art, but a way of living in contemporary Cameroon, foregrounding the population's ability to make a functional livelihood from what the rest of the world might otherwise discard.

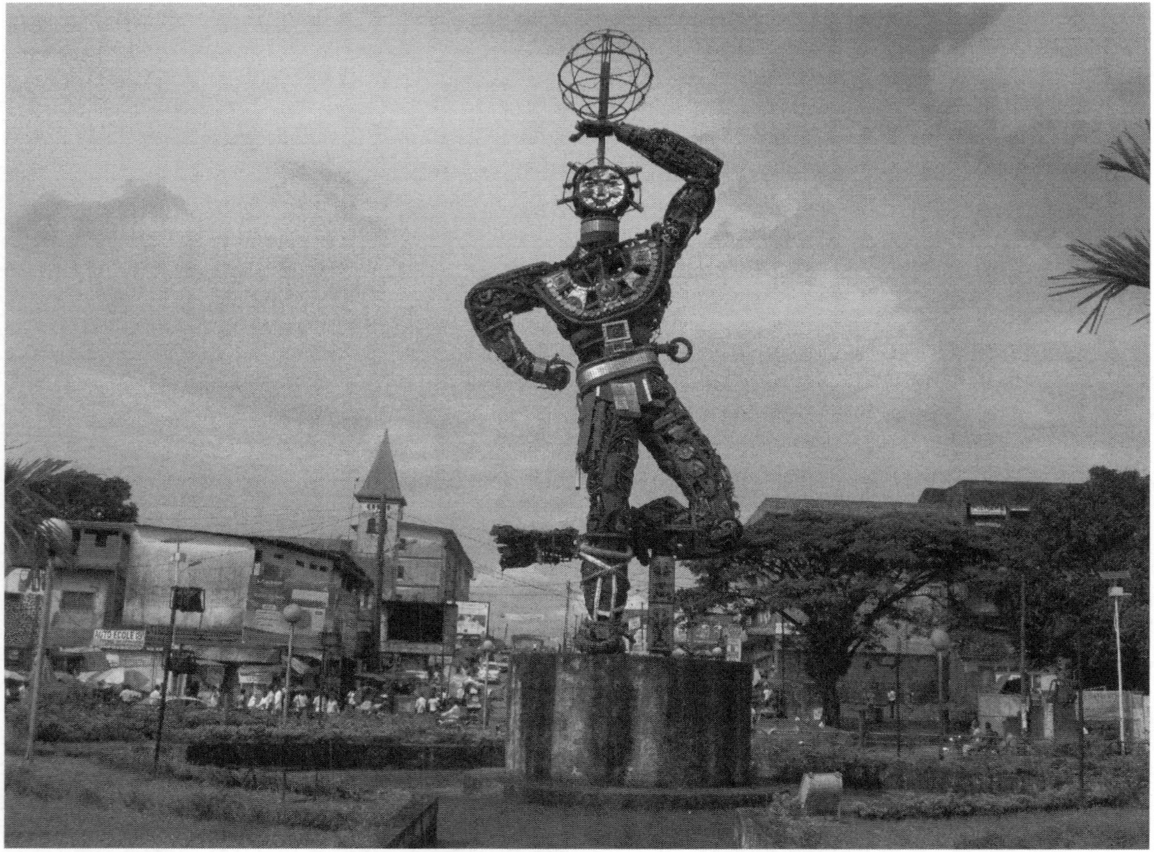

Figure 3.3 *La Nouvelle Liberté*, Joseph-Francis Sumégné, 1996 (source: Wikimedia Commons, photography by Theodorento1, 2013).

Today, in a world where we are more than ever surrounded by what we have discarded (take the replacement of fish in the ocean with plastic, for example), artists are often working with what is often called 'repurposed' material, pointing to transformation of waste and trash into art objects and installations. Jill Sigman, whose work to which I will now turn, emphasizes that working with discarded material is not limited to being merely an environmental project or an anti-littering campaign; rather, in line with *La Nouvelle Liberté*, it is a means to change the way we see and relate to the world around us.

Whereas Wentworth focuses on what can be done with the objects and the critical practices that already exist in urban space (Hawkins 2010), Sigman, like Sumégné, is more concerned with what objects mean, and on creating interventionist actions that promote new ways of being in the city. Hence, Sumégné's and Sigman's work complement Wentworth's, and add a fruitful perspective in terms of how artistic practice can make urban space available for criticism and questioning. As I will demonstrate in relation to Sigman's work, this is not only by making explicit the social significance of objects' observable and practical use, but also by understanding how objects are inserted into mental frameworks that relate to more ephemeral material practices such as destruction and dis/appearance. In the following, I will scrutinize Sigman's work in this regard.

Performing materiality with Jill Sigman

Jill Sigman is an American choreographer and transdisciplinary artist. She has been making dance and performance installations since the 1990s, and her work exists at the intersection of dance, visual art, and social practice. In 1998, the same year as she received her PhD in philosophy from Princeton University, she created *jill sigman/thinkdance* to ask questions through the medium of the body (Sigman 2007). Sigman's work has long blurred the boundaries between artistic forms, integrating trained dance with social experiment and community activism.

While Sigman has created a wide range of dance, theatre, and visual installations, what is particularly relevant for my research is her series *The Hut Project*, which is a series of huts made from found and repurposed materials. Each hut acts as a container for performance, tea, collaborations, and community dialogue. The huts are site-specific in the sense that they are made from things gathered in and around the place where they are built. As Sigman (2011) explains, she looks for things that are particularly interesting in a place and asks herself how gathering these things from the site can capture her experience of this place by becoming part of her work.

Sigman's practice revolves around how objects mediate our relationship to space:

> For example, in New York we have a problem with bed bugs, so when you see a mattress in the street you might go to the other side of the street not to go near it. So, it is performing its materiality so robustly that it changes our path through space.
>
> (personal interview, 06.09.2014)

However, Sigman also talks about how she as an artist can change people's pathways by having the objects 'perform' their materiality in different ways; by bringing these objects into a different space and context, she can give people different thoughts and experiences in relation to them. As she describes it:

> The huts are simultaneously stages, sculptures, and dwellings. Like ritual spaces, they are places where someone might have an experience they wouldn't have in daily life, and where that experience can be contained and understood through a special space created to hold it. Perhaps because they are made of waste, they are spaces outside of the consumer world and off the map of "normal" social transactions.
>
> (Sigman 2017, 4)

Furthermore, she emphasizes how the found materials, full of object histories, enable new kinds of connection and ways of relating as it is a connection that is not mediated by profit or product (Sigman 2017). She here points to a shared responsibility, in which the objects become avenues of a community's mutual responsiveness.

This approach is perhaps best illustrated through Sigman's *Hut #7* (2012), which was built in the northern part of Brooklyn, New York, an area having problems with chemical pollution, garbage, and illness due to a large amount of waste processing taking place there. The Newtown Creek canal, running through the area, is one of the most polluted bodies of water in the United States. When doing research on the Brooklyn neighbourhood by walking through the streets and talking with the residents, Sigman realized that the residents felt that the neighbourhood itself was treated like a garbage dump, and that garbage therefore was an important and problematic issue for them:

> Working with cast-off objects led me to many places and people who have also been 'cast off' or rendered invisible in various ways. [. . .] I saw and met people who also deal with waste, much more extensively than myself: homeless people, nomadic people, professional trash pickers and recyclers, dumpster divers, janitors, maintenance workers, guards, facilities personnel, and sanitation workers. It's often not a coincidence who is economically driven to take these jobs or live in these ways, and I was aware of my privilege in choosing to work with waste. It's also not surprising which places are "discarded" – whose homes and neighborhoods have been deemed disposable. Immigrants, people of color, the poor, and the otherwise marginalized have a much higher probability of living with toxicity and dealing with waste.
>
> (Sigman 2017, 11–12)

Sigman observed that a lot of the waste in the area was 'icky' as it had fallen off trucks and often had oil and construction residue on it. However, there was also waste coming from the luxurious condominiums located a bit further away, such as champagne bottles and various IKEA packaging. The contrast between the luxury waste and the industrial waste reflected in many ways the striking contradictions in the neighbourhood. The former Greenpoint Hospital, housing the community centre in which *Hut #7* was placed, was closed down over three decades ago. The part of the housing complex in which the community centre was located was in functional condition, whereas the hospital buildings at the other end of the block were still vacant and graffiti-covered with broken windows (Croghan 2014). In order to reflect what the residents were feeling about the neighbourhood, as well as the contradictions she observed in it, Sigman decided to use the material she found as a way to think about how the area felt toxic.

The idea of containment became important in this regard. Relating to the broader context of New York, *Hut #7* can here be seen as a response to the proliferation of policies and practices in the city aimed at (re)moving 'disturbing elements' such as marginalized people and urban waste to certain neighbourhoods where they are 'contained', so as to 'purify' other (more affluent) parts of urban space. Furthermore, Sigman wanted to treat the waste in ways so that people were no longer directly exposed to it. Hence, she wrapped the waste in different kinds of cloth, cardboard, and plastic wrapping materials that she also found in the area. This material was taken from the non-sites (in Smithson's sense of the word; see Chapter 1) of consumer culture, in other words the negative ready-mades of wrapping materials in which objects have been packed and shipped, thus displaying the discarded evidence of an infinite production of waste. The wrapped containers reminded Sigman of medicine bundles from indigenous cultures as well as body bags, and thus linked to the history of the building as a former hospital, as well as the present health issues in the area.

The primitive look of the containers established at once a distance and proximity to the objects. On one hand, the bundles could be seen as strange artefacts from a historically distant cult, enabling a critical analysis as if studying a foreign culture. On the other hand, the DIY aesthetics of the low-tech structure provided a sense of familiarity and closeness that suggested that anyone could engage with

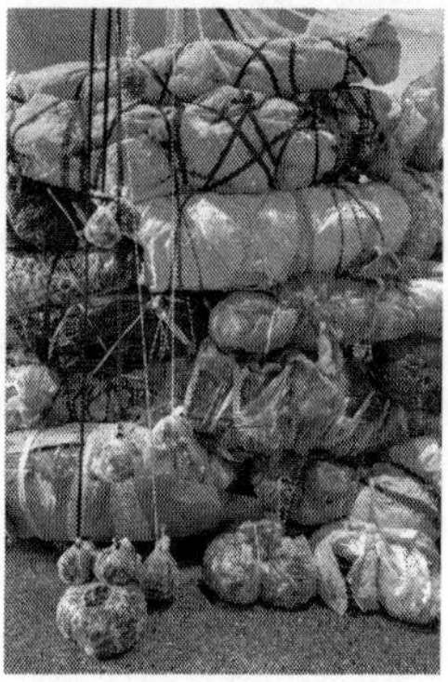

Figure 3.4 Hut #7, detail, Jill Sigman, 2012 (photo © Rafael Gamo, 2012).

the material on a personal level. Similarly, Edensor (2007) observes how tactile engagement with things usually consigned to landfill and dumps brings back some of the familiar sensations of his childhood, when he dwelt in dens and woods and ostensibly off-limits derelict houses. Sigman points to similar reactions from the visitors of the huts:

> Often, they tell me of [. . .] what they built when they were kids. In the beginning they are usually a little bit afraid of it, like: 'wow, what is that?' But then they start telling me about stuff they built when they were five-years-old, like: 'I built a fort!' So, they become a bit like kids. And sometimes they relate to the objects: 'oh, where did you find this? What is this?' [. . .] When we talk, sometimes it is about a certain wish for their neighbourhood, like what they would want to see if they could change something.
>
> (personal interview, 06.09.2014)

Sigman underlines how the hut changes the relationship between the artwork and the visitors; instead of the 'distanced look' of the visitor to a gallery evaluating the hut as an aesthetic structure, the visitors are invited into the hut by Sigman as if they are entering her home. The hut functions as a space in which things can be engaged with, moved, and strewn around, in contradiction to the engagement with things in more regulated realms, such as in an art gallery, where typically vision predominates, objects are beheld at a distance, seen as sacrosanct, and may not be meddled with. The objects that comprise the hut, on the other hand, have already been consigned to a category of waste, and thus can be engaged with by everybody as the waste does not belong to anybody. Accordingly, as Sigman observes (McLendon 2015), when visitors do things in the hut, they get a different relation to the space, their bodies, each other, and their ideas about what is possible. This is further emphasized by the scruffy DIY aesthetics of the hut, where the construction is supported by everyday materials such as cardboard and duct tape. This approach of 'using what is at hand' is, to Sigman, 'a way of resisting planned obsolescence in all fronts, unmaking a consumer culture' (McLendon 2015, 122).

Commenting on the artistic practice of the Swiss artist Thomas Hirschhorn, who applies a similar DIY aesthetic in his work, Anthony Gardner (2012) argues that this form of spontaneous and undesigned environment transmits Joseph Beuys' concern that everyone can be an artist – no training or specialist know-how is required to cut, screw, and paste together these forms of temporary structures. As Sigman puts it:

> The fact that [the visitors] enter the hut is really important; it is not an art object that they have to stay away from. Often, I make the entrance really low, or high, so that people have to get physical to get in there – crawl or bend to get in and then sit on the floor inside. In general, I want them to have a physical experience, which is not one of distance. This helps them feel that this is their space and not just a representation of space.
>
> (personal interview, 06.09.2014)

For Sigman, as a dancer and choreographer, the relationship between bodies and space, in terms of how the hut is experienced affectively, is crucial. The most important aspect of her work is to 'choreograph experience' through the web of actions and relationships that are initiated by her hut. At one point, Sigman made a three-hour performance, standing on the top of a ladder and holding an old car antenna outside of the hut. She describes the performance as 'a feat of physical endurance, an internal focusing, and a way of being connected to the viewers in a space that had the hut at its epicentre' (Sigman 2017, 19). After the performance, she spoke with a young man that had volunteered to run the bar. He told her that during the performance, he had found himself overwhelmed by emotion and had the urge to cry. According to Sigman, she could feel that charge when she was doing the performance.

As Claire Colebrook (2002, cited in Thompson 2009, 117) notes, 'art may well have meanings or messages but what makes art is not content but its affect, the sensible force or style through which it produces content'. The focus on producing feelings of embodied sensations such as joy, comfort, awe, or astonishment provides an alternative to the sole concentration on the utility of art, as criticized by Deutsche (1988). According to James Thompson (2009), professor of applied and social theatre, affect allows one to understand the substance and complexity of the work instead of reducing experiences to measurable indicators of effect such as an increased knowledge on particular issues, simplistic diagnosis for problem-solving, and participant involvement. As Sigman (2017) points out, by having the material 'perform differently' she aims to give 'the audience' different thoughts

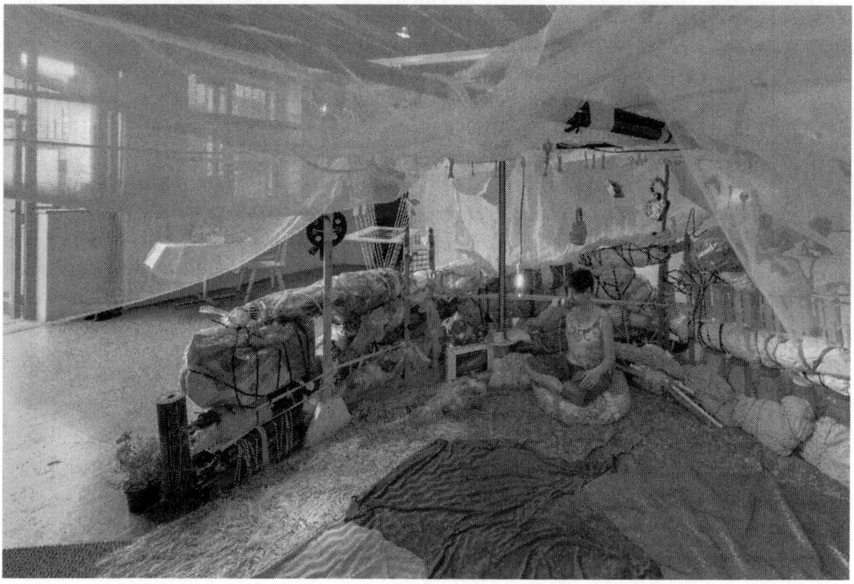

Figure 3.5 Hut #7, interior, Jill Sigman, 2012 (photo © Rafael Gamo, 2012).

and experiences relating to it, and in turn change how people think and experience the world: 'In the unfamiliar but intimate context of the huts, I witnessed people talking to others they didn't know, tasting things they wouldn't normally eat, and thinking thoughts they wouldn't often think' (p. 5).

Affect is connected both to a capacity for action and to a sense of aliveness that prompts a person's desire to connect and engage. Jill Bennet (2005, cited in Thompson 2009, 120) argues that affect produces real-time somatic experience that is no longer framed as representation. Accordingly, Sigman attempts to place the objects in an unmediated relationship with the viewer by 'freeing' them from the idea of any conscious artistic intervention. Sigman hence focuses on making her huts *real*, in terms of creating a form of inhabited space in which an immediate reflection on the world around us may take place.

Socially lived space

In order to create a form of *real* and unmediated space that is free of representation, Sigman performatively engages with the hut as part of her daily life: she grows plants inside it, as well as cooking and serving food, and sometimes sleeping there. Avoiding any forms of representation can be problematic as it risks neglecting the importance of engaging with the conceptual realms of space, as well as the material activities that produce them. Lefebvre's (1991) notion of 'socially lived space' provides helpful insights for understanding this importance.

Lefebvre observes 'an opposition established between our conception of space – abstract, mental and geometric – and our perception of space – concrete, material and physical' (cited in Elden 2004a, 96). In order to make progress in understanding space, he argues, we need to engage with their conceptual realms as well as the material activities within them; we need to grasp the concrete and the abstract together. Lefebvre thus provides us with a term between the poles of conception and perception, the notion of socially lived space. Space is here seen as real and imagined, depending on physical and mental constructs. Accordingly, socially lived space points to how the structures, signs, and codes of space produce and integrate with everyday life.

The notion of the *production* of a socially lived space as a perceived, conceived, and lived space is essential here, as it looks at the relation between politics and space. Production, as Stuart Elden (2004b) points out, does not solely refer to the Marxist notion of the strictly economic production of artefacts, but should also be seen as a larger philosophical concept in terms of the production of meaning, knowledge, discourse, and institutions. In other words, production is seen as the act of making all sorts of things beyond traditional material objects and even beyond the commodity. As Joseph (2002) observes, recognizing this diversity of production implies recognizing the diversity of subjects produced therein, be it proud shop owners, devoted neighbours, loyal social workers, gays and lesbians, and so on. Joseph (2002) here emphasizes the performativity of production, arguing for a view on production in which 'it is the potentially diverse performativity of productive activity and of products that matters' (p. 36). In line with Butler, Joseph (2002) defines performativity as 'the retroactively constituting enactment of discursive constraints and not the free act of a fully constituted subject' (pp. 33–34). Furthermore, it is not only the individual's identity that is expressed and formed through production, but also collectivity or social relations. Accordingly, the visitors to Sigman's hut were co-producers of the space through spatial practice, in terms of being invited to 'inhabit' the hut by entering it, eating the food that was served to them, and so on.

Representations of space and spaces of representation are equally important to Lefebvre's conceptual triad of socially lived space as is spatial practice: 'A park is *conceived*, designed and produced through labour, technology and institutions, but the meaning of the space, and the space itself, is adapted and transformed as it is *perceived* and *lived* by social actors and groups' (Elden 2004b, 191, original emphasis). When we remove representation from the equation, we risk enhancing the idea of space being natural and pre-existing, rather than in a constant process of becoming.

Conclusions

The cases that I have discussed in this chapter make explicit different ways in which urban space and the material environment are transformed and come into being in ongoing everyday practices. By working with and making explicit processes of meaning-making, artistic practice enables us to understand how people and things are not just in the world, but incorporated into practices in the world. Lefebvre's notion of a socially lived space, the theories of affordances, and surrealist approaches point to the different insights, conceptualizations, and theorizations that artistic practice can offer us in this regard. In line with the theories of affordances, Wentworth's work points to how objects can be seen as 'crystallizations' of human activities, inviting and constraining us to use them in certain ways, even if this use does not correspond to their intended function. The ability of materiality to influence those who interpret is further foregrounded by the work of Sumégné. Sigman takes this even further in the making of her huts, in which she makes explicit the function of objects as affective mediators between people and space. By removing the objects from their normal context and practical function, she aims to give people different thoughts and experiences in relation to them.

The focus of the artistic practices of Wentworth, Sumégné, and Sigman lies in the creation and conception of art, and less in the co-production of the artwork by the audience. The artists emphasize the importance of their individual discovery and collection of the objects they work with. Hence, whereas their practices illustrate how art may enable a form of investigation into how we can give meaning to the material world around us and our diverse experiences and encounters with it, these practices do less to elucidate aspects of co-production between artists and their non-artist collaborators. This is simply because their intention is not to produce socially engaged art as such. In order to demonstrate the strengths and weaknesses of co-producing artistic practice that directly engages with the urban environment, I will in the next chapter turn to artistic practices that have got explicit socially engaged aims and in which public participation plays a crucial part. In line with Lefebvre, I here emphasize an approach to participation in which inhabitants of the city increasingly come to manage the production of urban space themselves, in such a way that their own collective power is revealed to them and they recognize their role as stewards of the urban (Purcell 2014).

Bibliography

Benjamin, W. 1968. "The Work of Art in the Age of Mechanical Reproduction." In *Illuminations*, edited by H. Arendt, 217–252. New York: Harcourt Brace Jovanovich.

———. 1999. *The Arcades Project*. Cambridge, MA: Harvard University Press.

Bennett, J. 2005. *Empathic Visions: Affect, Trauma, and Contemporary Art*. Stanford, CA: Stanford University Press.

Boniver, T., Devlieger, L., Ghyoot, M., Gielen, M., Lasserre, B., Tamm, M., and Zitouni, B. 2010. *Usus/Usures: How Things Stand*. Brussels: Snel.

Bonnett, A. 1992. "Art, Ideology, and Everyday Space: Subversive Tendencies from Dada to Postmodernism." *Environment and Planning D: Society and Space* 10: 69–86.

Borden, I., Kerr, J., Rendell, J., and Pivaro, A. 2000. *The Unknown City: Contesting Architecture and Social Space*. Cambridge, MA: MIT Press.

Colebrook, C. 2002. *Gillez Deleuze*. London: Routledge.

Costall, A. 1995. "Socializing Affordances." *Theory & Psychology* 5 (4): 467–481.

———. 2012. "Canonical Affordances in Context." *AVANT* 3 (2): 85–93.

Croghan, L. 2014. "What Ever Happened to Greenpoint Hospital and Other Shut-Down Brooklyn Healthcare Institutions?" *Brooklyn Daily Eagle*, 28 October. Available at: www.brooklyneagle.com/articles/2014/10/29/what-ever-happened-greenpoint-hospital-and-other-shut-down-brooklyn-healthcare [accessed 10.06.2016].

DeSilvey, C. 2006. "Observed Decay: Telling Stories with Mutable Things." *Journal of Material Culture* 11 (3): 318–338.

Dezeuze, A. 2013. "Photography, Ways of Living, and Richard Wentworth's Making Do, Getting By." *Oxford Art Journal* 36 (2): 281–300.

Dezeuze, A. and Kelly, J. 2013. "Introduction: Involuntary Sculpture: Process, Photography and the Ephemeral Object." In *Found Sculpture and Photography from Surrealism to Contemporary Art*, edited by A. Dezeuze and J. Kelly, 1–12. Surrey: Ashgate.

Duncker, K. 1945. "On Problem Solving." *Psychological Monographs* 58 (5): 1–13.

Eastham, B. 2011. "Interview with Richard Wentworth." *The White Review*, 2. Available from www.thewhitereview.org/interviews/interview-with-richard-wentworth/ [accessed 08.10.2018].

Edensor, T. 2007. "Sensing the Ruin." *The Senses and Society* 2 (2): 217–232.

Elden, S. 2004a. "Between Marx and Heidegger: Politics, Philosophy and Lefebvre's 'The Production of Space.'" *Antipode* 36 (1): 86–105.

———. 2004b. *Understanding Henri Lefebvre*. London: Continuum.

Gardner, A. 2012. "De-Idealizing Democracy: On Thomas Hirschhorn's Postsocialist Projects." *Art Margins* 1 (1): 29–61.

Gibson, J. 1979. *The Ecological Approach to Visual Perception*. Boston, MA: Houghton Mifflin.

Gilloch, G. 2002. *Walter Benjamin: Critical Constellations*. Cambridge: Polity Press.

Hawkins, H. 2010. "Turn Your Trash into . . . Rubbish, Art and Politics: Richard Wentworth's Geographical Imagination." *Social & Cultural Geography* 11 (8): 805–827.

Highmore, B. 2002. *Everyday Life and Cultural Theory*. London: Routledge.

Hodder, I. and Hutson, S. 2003. *Reading the Past: Current Approaches to Interpretation in Archaeology*, 3rd edition. Cambridge: Cambridge University Press.
Joseph, M. 2002. *Against the Romance of Community*. Minneapolis, MN: University of Minnesota Press.
Knappett, C. 2004. "The Affordance of Things: A Post-Gibsonian Perspective on the Relationality of Mind and Matter." In *Rethinking Materiality: The Engagement of Mind with the Material World*, edited by E. DeMarrais, C. Gosden, and C. Renfrew, 43–51. Cambridge: MacDonald Institute Monographs.
Latham, A. 1999. "The Power of Distraction: Distraction, Tactility, and Habit in the Work of Walter Benjamin." *Environment and Planning D: Society and Space* 17: 451–473.
Lefebvre, H. 1991. *The Production of Space*, translated by D. Nicholson-Smith. Malden, MA: Blackwell.
Makongo, Y. 2018. "Giving Meaning to the City: Interview with Marilyn Douala Manga Bell." *Contemporary & Magazine*. Available at: www.contemporaryand.com/magazines/giving-meaning-to-the-city/ [accessed 08.10.2018].
Massey, D. 2005. *For Space*. London: Sage.
McLendon, M. 2015. "Jill Sigman." In *Re:Purposed*, edited by M. McLendon, 114–126. New York: Scala.
Mileaf, J. A. 2010. *Please Touch: Dada and Surrealist Objects after the Readymade*. Lebanon, NH: University Press of New England.
Pile, S. 2005. *Real Cities*. London: Sage.
Purcell, M. 2014. "Possible Worlds: Henri Lefebvre and the Right to the City." *Journal of Urban Affairs* 36 (1): 141–154.
Sheringham, M. 2006. *Everyday Life: Theories and Practices from Surrealism to the Present*. Oxford: Oxford University Press.
Sigman, J. 2007. *Jill Sigman/Thinkdance*. Available from www.thinkdance.org [accessed 08.10.2018].
———. 2011. *The Hut Project*. Available from www.youtube.com/watch?v=z-5eGjbf51w&list=UUiMeTp6djYDuXDlj4GNruHA&index=1&feature=plcp [accessed 08.10.2018].
———. 2017. "Paradise Declassified." In *Ten Huts*, edited by J. Sigman, 1–24. Middletown, CT: Wesleyan University Press.
Simone, A. 2004. *For the City Yet to Come*. Durham, NC: Duke University Press.
Tambi, M. D. 2015. "Economic Growth, Crisis and Recovery in Cameroon: A Literature Review." *The International Journal of Industrial Distribution & Business* 6 (1): 5–15.
Thompson, J. 2009. *Performance Affects*. London: Palgrave Macmillan.
Warner, M. 1993. *Richard Wentworth*. London: Thames & Hudson.
Whatmore, S. 1999. "Hybrid Geographies: Rethinking the 'Human' in Human Geography." In *Human Geography Toay*, edited by D. Massey, J. Allen, and P. Sarre, 22–39. Cambridge: Polity Press.
Wood, M. and Flinders, M. 2014. "Rethinking Depoliticisation: Beyond the Governmental." *Policy and Politics* 42 (2): 151–170.
Woodward, C. 2002. *In Ruins*. London: Vintage.

4 Mobilizing communities

Group Material and the politics of representation

Group Material is the name of an artistic collective that was active in New York from 1976 to 1996. The Group was founded at a time when New York was on the brink of bankruptcy. City residents lived precariously and in deprivation, forced to endorse a reality they felt was imposed on them by executive powers (Ashford 2010). The new influx of wealth into the city during the 1980s, following the boom on Wall Street, paved the way for gentrification as inner-city rents shot upwards. This was the scene for Deutsche's critical writings on uneven urban developments, in which she observes how New York became increasingly functional and alienated from the social life of its inhabitants (Deutsche 1988).

Group Material's artistic aim of mobilizing communities in changing their circumstances can be seen in the light of these developments and what the Group identifies as 'the emergence of a right-wing culture of physical control and spectacularized consistency' (Ashford 2010, 224). Doug Ashford, one of Group Material's members, points to how the collective modelled the revolutionary counterculture of the 1960s in their desire for political change based on the imperatives of social organizing and the methodological sensibilities of artists (Ashford 2010). The name 'Group Material' here signifies a 'materialist' approach, in terms of working as a collective against individual art practices, and aiming to connect art's production with its reception (Green 2011). The Group's projects are claimed to foreshadow 'the social turn' (Bishop 2006) in contemporary artistic practice, as well as 'relational' (Bourriaud 2002), 'context' (Lippard and Chandler 1968), and 'dialogical' (Kester 1999) practices. However, the Group is, as Alison Green (2011) points out, curiously absent from recent discussions on contemporary socially engaged art.

The work of Group Material raises several important issues concerning forms of production, community participation, and the idea of social change through art. The Group made considerable efforts to mobilize a 'dialectical approach to reality through the means of art' (Avgikos 1995, 92). This effort was predicated on the synthesis of two separate and distinct models of social space: the gallery and the neighbourhood. Parallels can here be drawn to Lefebvre's programme for radical change. As Elden (2004) points out in his analysis of Lefebvre's work, Lefebvre makes it clear that a revolution cannot just hope to change the political personnel or institutions; it must also change the everyday in order to foster 'the art of living' (p. 118). To do this, one needs to critique the trivial through the exceptional, and at the same time critique the exceptional with the trivial. This implies a criticism of the elite by the mass, and of dreams, art, and poetry by reality. Elden (2004) observes that, to Lefebvre, this is not an either/or choice or a one-way critique, but a dialectically driven process. In line with this approach, Group Material insisted on rethinking art from the ground up, so that art would become relevant not only to the lives of the Group's members, but to those disenfranchised audiences, such as working people and non-art professionals from all walks of life, with whom they identified.

People's Choice

When moving into their storefront art space on Manhattan's Lower East Side in 1981, at that time an economically depressed and predominantly Hispanic neighbourhood, Group Material was determined to ground its grassroots practice and programming in relation to the neighbourhood community. The neighbourhood was one of the last significant low-rent districts in Manhattan, south of Central Park, and quickly became a hotspot for young artists seeking cheap housing, studio space, and a new sense of community (Bowler and McBurney 1991). Well aware of their role as gentrifiers in the area, Group Material wanted to make art 'in concert with communities in crisis with a direct intention to change things to positive effect' (Rollins 2010, 217). As Tim Rollins (2010), another member of the Group, emphasizes, 'We wanted to be communitarian (not Communist), not only community-based but also community-engaged, connecting what happens inside the exhibition space/headquarters with social life on the street and neighbourhood just outside the doors' (p. 218).

Accordingly, the Group went door to door in the neighbourhood with letters inviting the residents to contribute personal possessions for their next exhibition called *People's Choice*. The possessions should be 'things that might not usually find their way into an art gallery' (letter to residents, printed in Ault 2010), and that the residents had a personal relation to, whether it was things that they found particularly beautiful, things that could communicate to others, or that meant something special for them, their families, or their friends. The residents were encouraged to write down the story of the object on labels to be displayed together with it. This way, the Group put emphasis on a 'democratic attitude' in which the display of the objects supported the set of meanings already granted by the owners of the objects, rather than by the artists. In reviewing the show, the artist Thomas Lawson commented, 'The value of these artifacts lay precisely in their sentimentality, a quality that is absent from most artwork that strives to mean something to a general audience' (cited in Green 2011, 19).

People responded generously, and the exhibition was filled with all kinds of artefacts: china dolls, family photographs, posters, a Pez candy dispenser collection, religious imagery, tchotchkes, folk art, and so on. Group Material points out that the exhibition was intended to be a forum in which multiple points of view were represented in a variety of styles and methods, mirroring the various forms of representation that structure our understanding of culture (Avgikos 1995, 116). This position can be seen as an immediate legacy of the surrealist movement: that reality is never directly accessible, but exists only in the multiple codes that represent it (Clifford 1981). The 'will to order' is here seriously undermined, in favour of surrealism's technique of a montage that offers a bombardment of materials that resist narrative resolution (Highmore 2002, 81). As Group Material states, 'we are not interested in making definite evaluations or declarative statements, but in creating situations that offer our chosen subject as a complex and open-ended issue' (cited in Avgikos 1995, 116). This approach can, in line with Joseph's (2002) observation mentioned earlier, be criticized for emphasizing the insufficiency, oppressiveness, and homogenizing character of any realist (political or cultural) representation, and enhance the idea of the unrepresentability of space. However, by facilitating different representational modes, *People's Choice* aimed at opening up the possibility of disarticulation, in which the disruption of one element by another challenges the authority of any one representational mode, and thus allows the problematization of representation itself (Highmore 2002).

Negotiating the value of art

Group Material wanted to change the world materially by moving beyond mere observation to interpretation and intervention: 'We encourage greater audience participation through interpretation' (cited in Avgikos 1995, 116). By having the

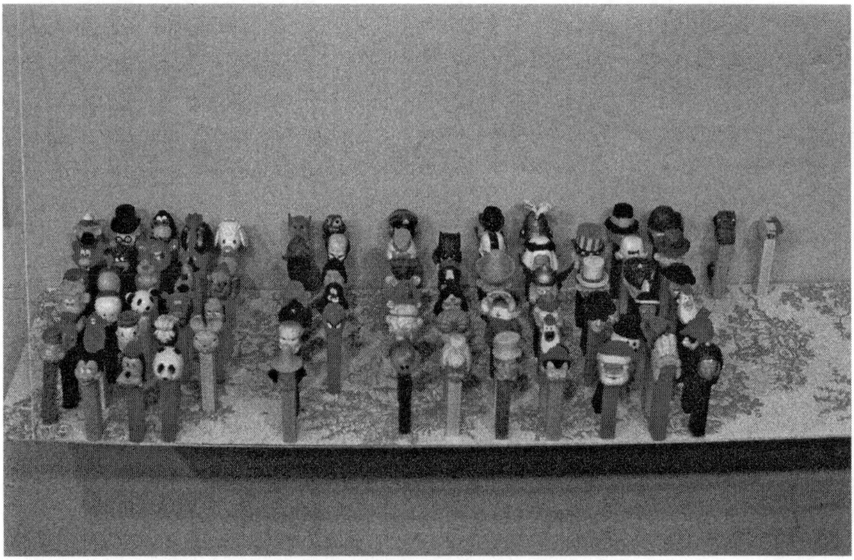

Figure 4.1 Pez dispensers, exhibition detail, *The People's Choice (Arroz con Mango)*, Group Material, 1981 (© Group Material, 1981).

neighbouring community provide the 'artworks', instead of the artists themselves, the Group put focus on the very production of art, asking, 'Who makes it, where is it, how does it get constituted?' (Ashford and Mohaiemen 2008, 40). The Group refused to define the existence and function of art as independent from the work, in terms of the conceptual processes, the physical labour, and the collaborative efforts required to produce it. Rather, the Group was in support of 'an ideology that values process as product, subject as object – and work as art' (Avgikos 1995, 90).

Group Material's focus on process as product and work as art can be seen in relation to a critique of the commodification of social life by modern capitalism. This critique relates to the emergence of art objects as commodities to be bought, sold, and appreciated for their purely aesthetic qualities (Bonnett 1992). By drawing attention to marginalized and second-hand objects from the people's homes, Group Material eschewed exchange value in favour of more idiosyncratic markers of worth. The 'base' of social reproduction was here seen not only as economic, but also as relating to non-commodity social forms such as cultural, bodily, and subjective relations (Rasmussen 2004). Accordingly, Group Material insisted on the substantive value of work as equal, rather than subordinate, to art. Hence, the Group expanded the social terrain of the artwork, determining its value not through the social process of inscribing the object as useful in its particulars, but through the social process of production itself.

The challenge of mobilizing communities

Group Material used every means at their disposal to precipitate the vital exchange between the gallery and the neighbourhood, and hence reach out to 'the people'. In addition to exhibitions such as *People's Choice*, they attempted to mobilize the community through parties, potlucks, art classes for kids, movie screenings, and so on. This way, they wanted to expand the boundaries of art to address issues shaped by the special character of the neighbourhood and the lives of its inhabitants. However, the collaboration between Group Material and the residents stalled at the most basic level. Their gallery never developed much beyond an art space that was operated and curated by a collective of young artists eager to mobilize 'the people' and give them art (Avgikos 1995). In a press release announcing the closing of the gallery one year after its opening, Group Material states the following:

> The mistake was obvious. Just like the alternative spaces we had set out to criticize, here we were sitting on 13th Street, waiting for everyone to rush down and see our shows instead of taking the initiative ourselves for mobilizing into more public areas.
>
> (cited in Rollins 1982, 22)

An easy critique to make in this regard is that Group Material simply colonized the neighbourhood and that the events they provided for the residents were perceived as nothing more than free entertainment by friendly 'outsiders'. However, this is a simplification. Lefebvre's notion of a socially lived space enables a more nuanced critique: Group Material's failure is that they neglected the fact that space is as dependent on conceptual realms as it is on material activities. In spite of hosting a series of community events, the gallery space did not function beyond the mental conventions incumbent to it. Unlike the church or the school, the gallery did not correspond to a traditional community space, but rather replicated a system of display and distribution analogous to commerce and high culture. As Group Material put it themselves:

> We've learned that the notion of alternative space isn't only politically phony and aesthetically naive - it can also be diabolical. It is impossible to create a radical and innovative art if this work is anchored in one special gallery location. Art can have the most political content and right-on form, but the stuff just hangs there silent unless its means of distribution make political sense as well.
>
> (cited in Avgikos 1995, 99)

The gallery did not become a community hotbed for political protest that would spawn locally organized campaigns against rising housing prices, poor educational facilities, and lack of political representation. Instead, the gallery remained an artists' space and an example of the gentrification that took place in the neighbourhood. Hence, whereas Group Material had attempted to combat the alienating effects of capitalism from 'outside' the capitalist mode of production, the Group was instead, in fact, a product of it. Their oppositional stances could not escape the system that they wanted to challenge.

Group Material's politics of representation, then, backfired. By resisting narrative resolution and focusing on a montage that offered a bombardment of materials, the group wanted to oppose the insufficiency, oppressiveness, and homogenizing character of any realist representation of a social group. Instead, they promoted a liberatory exteriority of presumably free, and thus less hegemonized, repressive, possessable, forms of signification. However, in doing so, the Group went into the trap of neglecting the relation between representation and production. As Joseph (2002) stresses, a critique of (capitalist) production needs to look at representation and the performativity of production, as well as at social and material formations, in order to recognize the opportunities as well as the constraints available within production. This brings me to the practice of the world-renowned Chinese artist Ai Weiwei. While Ai is perhaps best known for his overt political dissidence against the Chinese authoritarian government, the social and political engagement of his work foregrounds the complex relations between art and 'the system', as well as between artwork and participants in the context of collective art-making.

Ai Weiwei and the provocation of publicness

The first decade of the twenty-first century in mainland China has witnessed a growing interest and undertaking of socially engaged art projects focusing on exposing the injustices created by an authoritarian regime, raising awareness of social problems produced by the logic of neoliberal urbanization, and fostering community-oriented interactions and participation (Wang 2017). This growing interest can be seen as an unintended consequence of the 'opening up policy' introduced by Deng Xiaoping in 1978 in order to liberalize markets and boost economic growth as well as technical innovation (Callahan 2014). While the aim was economic reform, the outcome of this policy has been less control over daily life, more space for discussion, and accordingly the opening up of a wide variety of cultural opportunities. As Bo (2009) observes, 'with the establishment of private properties, citizens increasingly resort to a rights-based rhetoric to defend their interests' (p. 82), resulting in the rise of a strong bottom-up desire for civil society, in turn characterized by a rapid growth of overt activism and non-profit organizations. Furthermore, while the media is still suffering from state censorship, the Internet has made it easier for citizens to voice and circulate their opinions.

Following these developments, the notion of 'publicness' has, over the past decade, received increased attention in Chinese art discourse (Bo 2009). While the concept of 'the public sphere', defined here as a stable realm of democratic public opinion-making, is very much rooted in European history, Bo (2012) argues that struggles for such a realm have existed and continue to be waged in contemporary China. Despite the economic reforms, much of the political system remains unchanged. The party-state remains a totalitarian power above the law, and freedom of expression, association, and publication still lacks institutional guarantees. The question, then, is, as Callahan (2014) puts it:

> How to strive for publicness when the state is not subordinated to public control, when citizens' rights to speak, associate, and publish are not protected by an independent legal system, and when the media is not free from state censorship.
>
> (p. 75)

Given the lack of basic institutional guarantees, attempts at realizing 'publicness' often appear as provocative and rebellious, as we will see in relation to the socially engaged art of Ai Weiwei.

Ai Weiwei has become somewhat of a global political art figure, hailed for his human rights activism by some and loathed for playing to the dissident-hungry Western media by others. As Callahan (2014) points out, Ai's work is full of contradictions and tensions that offer valuable insights for understanding the relation between socially engaged art and public engagement. While the political dynamics in China are central for understanding this relation, his practice raises general and important questions about the production of socially engaged art, in line with Group Material's questioning of how art is made, where it is, and how it gets constituted. Ai's *Sunflower Seeds* installation at London's Tate Modern art gallery in October 2010 is a particularly good example in this regard as it put the tensions between conceptual processes and material activities in the social process of art production into sharp relief.

The *Sunflower Seeds* installation covered the Turbine Hall in 100 million real-sized porcelain seeds, hand-painted and manufactured by 1,600 craftspeople in the capital of China's ceramics industry, Jingdezhen. In his brilliant discussion of the collaborative efforts that underpinned the installation, Bo (2012) points to a central disconnect between the material production of the installation and the discursive sphere in which its meaning-making was produced. While Western critics celebrated the work for its political potential in terms of opposing the Chinese regime, as well as for rescuing the workers painting the seeds from unemployment and for supporting traditional craft production, Chinese critics focused on the work's relation to global capitalism and geopolitics. This critique questioned the work's relevance for the sociopolitical conditions of contemporary China, pointing to how it replicated the doings of the current government in terms of high investment in a massive production made for export by cheap labour, often at the expense of the environment.

As Bo observes, when exhibited in the UK, the voice and opinions of the workers themselves were curiously absent from any documentation of the work's

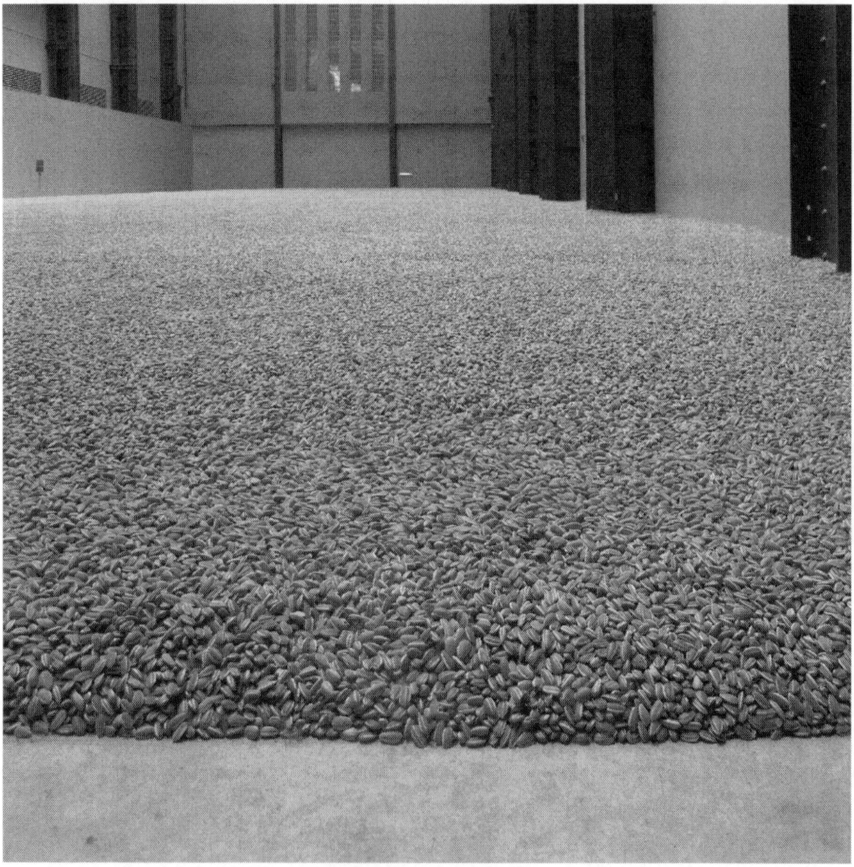

Figure 4.2 Installation view, Turbine Hall/Tate Modern, *Sunflower Seeds*, Ai Weiwei, 2011 (source: Wikimedia Commons).

production. The shots in the Tate documentary video are of the workers working silently, and when engaged in conversations with the artist himself, the discussion concentrates solely on the labour they are carrying out. Sensitive aspects such as the financial transactions between the artist and the factory owners were hidden from view, and not a single worker was seen as wearing a mask, even though it was considered a health risk to let the Western audience walk on top of the seeds due to the dust arising from them. The parts of the documentary in which Ai talks about the aesthetics and politics of the work is in English and directed towards the visitors of the installation. Hence, while the Tate audience were invited to reflect and join a public conversation, the workers were relegated to the private sphere of object production, not provided the opportunity to take on the role as active citizens participating in the public sphere of meaning production.

Similar to Group Material's *People's Choice*, there is a disconnect here between the material activities and the conceptual realms in which *Sunflower Seeds* was produced. The artist is the only link between the two realms. The work therefore only represents the expression of a single artist, while the Chinese workers remain voiceless. While the Chinese word for worker – *gongren* – had revolutionary connotations as the leading class towards communism in the state-socialist era, neoliberal capitalism has replaced the image of a strong, proud, and vocal worker with a young, obedient, and voiceless individual whose only social relation is to that of capital (Bo 2012). In spite of being part of producing the work, the participation of Chinese workers very much replicated this latter image of the silent and obedient workers. This form of participation, then, did not constitute any form of critical expression in the public sphere. The question, then, is how socially engaged art in China can enable participants to act not only as workers, but as citizens contributing to the production of a public sphere in China. I will now turn to Ai's project *Nian* to answer this question.

Nian

Nian, meaning both to commemorate and to read in Chinese, is a 220-minute-long collaborative sound work to which over 3,000 volunteers contributed their voice. One by one, the names of over 5,000 children killed in the 2008 Sichuan earthquake are read out loud in different dialects, with different pitches and intonations, and with various sound qualities. To make the artwork, Ai, together with other activists, managed to compile a list of the names of the children and sent out a Twitter message encouraging his followers to read out loud one or a few names, record the reading, and email him the recordings. These were then compiled into a single file and distributed via the Internet (Bo 2012).

The work can be seen as a response to the lack of responsibility taken by the government in relation to the fact that the children were killed as school buildings collapsed over them when they were in their classrooms. Parents questioned the poor construction of the school buildings and demanded that the local government investigate matters of official corruption linked to the construction of the buildings. The government not only denied such wrongdoings, but also prevented parents and activists from pursuing further investigations (Wang 2017). However, Ai and a couple of other activists refused to let the issue go, and initiated their own investigations, what Ai called *Citizens' Investigations*. Bo (2012) points out how the word for 'investigation' – *diaocha* – was an official word appropriated from the political vocabulary of the Mao era and representing a standard practice of the Chinese Communist Party. By linking this word to the notion of 'citizen', Ai turned the relation around; insisting on the right of the people to hold the state responsible, rather than the other way around.

According to Bo (2012), *Nian* posed a fundamental challenge to China's authoritarian system as it highlighted the citizens' right to engage in activities independent and critical of the state. Furthermore, the people's participation in the artwork shook the ideas of solidarity promoted by the Chinese state. The idea of solidarity as a mutual bond within a given social group or community, as discussed in Chapter 2, fits well with a totalitarian regime in which relationality is based on each individual's predefined and static role in the system. In other words, the parents' private grievances and struggle for further investigation by the government constituted a form of solidarity that posed less threat to the system because it was based on the pre-existing kinship of family – a solidarity with those similar to ourselves. *Nian*, however, enacted another form of solidarity, a solidarity based on difference. The readers of the names in the sound work declared themselves in solidarity with total strangers that they might have nothing in common with. As discussed in Chapter 2, this form of solidarity requires that you are able to partly dis-identify yourself from your own position. By doing this, you acknowledge that your own position in society is not given and can always change. This kind of contingency poses a threat to the authoritarian system because it represents a form of 'stranger-relationality' (Warner 2002) that denies any predefined and organizational positioning of a subject within a given system. Since the Mao era, the Chinese state had worked hard to obliterate this kind of contingency as it clearly posed a threat to the state's stringent control, in which a person's relation to the state trumped all other types of relations (Bo 2012). While the economic reforms have resulted in some more freedom and flexibility, interaction between people is now following individual interests that are oriented around transactions rather than relationality: What can I gain from my interaction with you?

Transient relationalities

Nian countered these understandings of relationality as being static and/or profit-driven with establishing a relationality that was relative, transient, lacking institutional basis, and situated within the project only. In doing so, the artwork opposed the dominant discursive arena in which publics are addressed as a stable and given entity and the mode of address is a rational-critical dialogue focusing on replicating opinions and persuasions in a uniform and replicable way (Warner 2002). The sound clips that make up *Nian* follow a different protocol of public expression. Instead of a uniform, stylized, and summarizable expression, the different poetic and textual qualities of the sound clips are foregrounded – the voices, dialects, pronunciation, background sounds, sound quality, and so on vary significantly. This linguistic fragmentation emphasizes the affective and performative qualities of the participating voices, opening up what Kanngieser (2012) terms 'an acoustic politics of the voice' (p. 337). Central to this form of politics is the idea that the way we speak and listen is political, and that voice and space co-create each other. As explained by LaBelle (2010, cited in Kanngieser 2012, 1):

> Sound operates by forming links, groupings, and conjunctions that accentuate individual identity as a relational project. The flows of surrounding sonority can be heard to weave an individual into a larger social fabric, filling relations with local sound, sonic culture, auditory memories, and the noises that move between, contributing to the making of shared spaces. This associative and connective process of sound comes to reconfigure the spatial distinctions of inside and outside, to foster confrontations between one and another, and to infuse language with degrees of intimacy.

The voice is here seen as more than a conduit for the transfer of information. By focusing on the extralinguistic elements of communication, *Nian* emphasizes the many different soundings, gestures, and affective transmissions that make up our different relations. The accentless narration of the state broadcasts are here opposed with intimate and personal voices made public. Accordingly, *Nian* enables a collective platform and discursive space for participants to identify and demarcate themselves as political subjects. The creative production of the sound work and the process of meaning-making are here closely linked; the participants forge relationships with strangers through their reading of the children's names, while at the same time making that stranger-relationality visible.

The distributed architecture of the Internet played a strategic role in this process as it offered the possibility to evade state censorship, to engage strangers, and to mobilize a critical discourse. The lateral form of distribution enabled by the Internet opposed the hierarchical organization of the state in which information is spread vertically, from the top down. In other words, the medium of the Internet becomes the message as it does not merely represent a vehicle for ideals of responsiveness, openness, and active coordination among strangers, but also embodies them. However, as mentioned in Chapter 2, social media can be critiqued for extending social engagement into the private sphere of individuals at the expense of group dynamics, offline public action, and grassroots organization. Strafella and Berg (2015) furthermore argue that Ai caters for a self-selected audience through his digital activism. The Chinese population has to employ software such as VPNs to circumvent the Great Firewall and access forbidden content such as Twitter; hence, the Chinese Twitter population may be more likely than other citizens to share Ai's yearning for freedom of expression. And finally, as Dean (2010) points out, digital media's invitation to participate by commenting and sharing, rather than reinforcing said values and aiding political mobilization, is easily co-opted by capitalism's flow of information and entertainment.

Because the state is so strong in China, it is particularly easy to frame socially engaged art as being in either radical opposition to or co-opted by the state. Callahan (2014) argues for the need to shift away from 'seeing politics in terms of ideas and audiences (and agency and structure), to concentrate on the emergence of civil society as a network of social connections that both includes and excludes the state' (p. 914). To Callahan, artists should be seen as liminal figures who sometimes work with the state and sometimes against it, but always for what they see as for the good of China and humanity more generally. Shifting focus to contemporary Europe, I will in the next section discuss the socially engaged art of the Dutch artist Jeanne van Heeswijk, who deliberately occupies this liminal space between autonomy and instrumentalization in relation to socially engaged art.

Creating collective narratives with Jeanne van Heeswijk

Jeanne van Heeswijk emphasizes that her socially engaged practice centres on participation in the system in terms of working from within, not from without. This 'within' refers not only to her positioning within 'the system', but also to her positioning within the neighbourhood. In this context, van Heeswijk sees her socially engaged art as a framework in which collective processes may occur (van Heeswijk 2012). Inspired by a particular event, cultural context, or intractable social problem, she involves neighbours and community members in the planning and realization of a given project. Often, she works with long-scale projects, embedding herself as an active citizen in communities for an extended period of time, involving hundreds of participants in performative actions, meetings, discussions, seminars, and other forms of organizing.

Whereas Group Material occupied a productive role in the culture wars in the United States of the 1980s and 1990s (Kwon 1997), in a context in which both progressives and neoconservatives learned how to mobilize 'communities', van Heeswijk started working with socially engaged projects in Central Europe in the 2000s – at a time when the question of what the arts can do for society was highly popular with European governments and art establishments.

Van Heeswijk argues that an autonomous art, positioned outside society, is not possible, and critiques the ways that this idea creates an unproductive binary between autonomy and instrumentalization: As mentioned in the introduction to this part, this binary implies that artists are inevitably instrumentalized when collaborating with various local partners, and ignores how these collaborations may enable artists to be instruments for change by, for example, enable a (re)occupation of the places in which we live. Accordingly, van Heeswijk argues that it is not enough to address local issues through debates; one also has to mobilize existing, local, physical, and sociocultural capital:

> For instance, in the case of the work in the [Afrikaander] market in Rotterdam, we identified a number of market sellers, people who might not live in the area, but who nevertheless through their roles have specific knowledge of the place. My interest lies in how they can – through forms of interaction and conflict with other 'experts on location' – become agents of change. And this can happen only if you are part of that process yourself.
>
> (cited in Gandolfi 2010, 137)

Van Heeswijk's view corresponds with a somewhat polemic statement by the Belgian writer Stefan Hertmans (2011), critiquing the so-called 'commitment' of contemporary art:

> I acknowledge immediately that it is noble, good and virtuous to stage plays that encourage the integration of non-European Union immigrants, but to be honest I work more effectively at integration by buying meat at a Turkish butcher or setting up a neighbourhood group with my Moroccan neighbour.
>
> (p. 57)

Van Heeswijk has worked extensively at organizing and establishing community groups that encourage local people to 'take matters into their own hands' (van Heeswijk 2011). For example, with the *2Up2Down/Homebaked* project in Anfield/Breckfield, van Heeswijk initiated and worked for over four years on making structural and physical change in the neighbourhood. As a result, the community worked with architects on remodelling a block of empty property to provide an affordable housing scheme, bakery shop, and kitchen, as well as meeting and project spaces, with the expressed needs of the residents in mind. Furthermore, a Homebaked Community Land Trust was set up in order to enable collective ownership of the properties and to reopen the bakery as a social enterprise (Stanhope 2015).

Whereas *Homebaked*, at first glance, primarily relates to a space conceived as constituted through labour and institutions, van Heeswijk stresses the importance of also working artistically with the meaning of space. She refers to what she calls 'collective narratives', which allude to the reimagining of cities by helping people to shape an idea of how they want to live together and what they would like their cities to be like. This process has much in common with theatre, defined by Alan Read (1993) as 'a process of building between performers and their constituencies which employs the medium of images to convey feeling and meaning' (p. 5). Read points to how theatre provides one of the most valuable means through which communities understand themselves as a collectivity and become understood by others. Drawing on Benedict Anderson's (1991) work on 'imagined communities', Read (1993) argues that images and the imagination are 'the means through which the material needs of communities and their mental aspirations form and disperse, at each point allowing a break with past dogmas and the expression of intangible possibilities' (p. 101). Theatre is understood as being constituted by both material and mental images, and can, according to Read, be seen in relation to the social space envisioned by Soja. This social space is defined as partly a mental space projected onto reality and given symbolic meaning (Read 1993, 158), and thus has much in common with Lefebvre's socially lived space.

The 'collective narratives' produced in van Heeswijk's practice are used as 'performative' bases to articulate these social and socially lived spaces. The notion of the performative is here understood as:

> A gesture or intervention that tries to create a space in which a diversity of individual projections might confront each other versus the projection of 'the perfect life/good' by project/real estate developers and local governments, who are each time trying to abstract and colonize space in order to create a different image of the city. In other words, we have to find a different way to use existing and new cultural capital.
>
> (van Heeswijk 2012, 80)

I will now discuss one of van Heeswijk's projects, *Wastelanders – Occupation of Transformation*, in which the focus on creating collective narratives is particularly explicit, to further scrutinize how socially engaged artistic practice may help articulate the relation between the material world and urban imaginaries.

Wastelanders – Occupation of Transformation

The project *Wastelanders – Occupation of Transformation* was initiated as part of the 2012 edition of the Parckdesign biennale in Brussels. The biennale was organized by Brussels Environments in cooperation with Architecture Workroom Brussels. The focus of the biennale was on 'the reorganisation of green space', with a particular focus on 'the transformation of undeveloped areas of Brussels' (Architecture Workroom Brussels 2012). These so-called 'undeveloped areas' were Anderlecht, Molenbeek, and Cureghem, three multicultural and deprived areas with a high crime rate and dominated by garbage and wastelands. The city of Brussels wanted these areas to be part of 'Brussels Green City', and used the biennale as a way to generate ideas for how they could become more green and sustainable. Artists, landscape developers, architects, and designers were invited to design and stage small-scale interventions in the abandoned and disused spaces there. As Architecture Workroom Brussels (2012) explains, the ambition was 'to closely adapt the physical interventions to the socio-economic reality of the neighbourhoods in question'. However, van Heeswijk points to a significant discrepancy in this regard: Anderlecht, for example, has one of the biggest second-hand car industries in Europe, on which a majority of the residents are dependent. Hence, focusing on making Anderlecht greener could be seen as a strategy for the city to remove the car industry and upgrade the area through a gentrification process that could eventually displace its original residents and make room for a more affluent middle class.

The relationship between material design and the everyday use of urban space is an important context for *Wastelanders*. Mattias Kärrholm's (2007) conceptualization of 'territoriality' (p. 438) is useful to understand this relationship and the power relations that it implies. Territoriality is here defined as 'a specific kind of power that uses space as its medium' (p. 439). The 'territory' is not static, but a mobile phenomenon that is continually re-marked and re-presented in different ways. Territoriality depends on control, socialized behaviours, and artefacts to remain effective. Borders and control are thus the result of territorialization, rather than vice versa. In order, then, to understand and work with the relation between material design and everyday use, it is not sufficient solely to criticize the strategies or reasons underpinning the construction of territorial order; one must also look at territoriality *in actu* in terms of the functions it actually serves.

Instead of openly criticizing what she observed as a hidden agenda for the Parckdesign biennale, van Heeswijk focused on articulating and making explicit the diverse narratives and functions that were already present in Anderlecht. She walked around the neighbourhood and asked people to share their stories about the area and their use of its public space. The stories were manifold. A group of women expressed their wish for a stone barbeque and comfortable benches in the park now mostly used by dogs, so that they can use the benches to gather their whole family. Charles told about how he sees the area as a large vegetable garden that one has to maintain properly; he uses derelict green spaces to plant and harvest vegetables that end up in his casserole. Simon works with young people in the area to manage and clean the playgrounds that otherwise are too polluted to use. Ilies and Moustafa use their car as a mobile studio to rehearse their raps; they park in deserted places in the neighbourhood so that they can produce music at full volume without disturbing anyone (Stanhope 2015).

Collective narratives as means for territorial production

The narratives collected in *Wastelanders* illustrate what Kärrholm (2007) terms 'territorial production' (p. 440), pointing to the possibility of cooperation in which several territorial strategies can be produced at the same time and place. Territorial production provides an alternative to singular territorial domination based on predetermined power relations in terms of, for example, who is in charge and who is not. Instead, territorial production presupposes power as a network of different actors, artefacts, persons, rules of conduct, and so forth, suggesting that territoriality is an altogether mobile and dynamic phenomenon. This opens up for investigation the meaning of materiality and artefacts through the roles they play in different territorial networks. Furthermore, it opens up possibilities for the investigation of the cooperation and coexistence of different territorial productions and powers in urban space.

Van Heeswijk filmed the 'collective narratives' of Anderlecht and showed them in a 'trash taxi' that transported visitors of Parckdesign from location to location around the neighbourhood for more than three months. The taxi was driven by local taxi drivers, sharing personal stories and observations from the area as they drove around. In exchange for a ride, passengers were asked to help collect the waste on the route in baskets on the car roof.

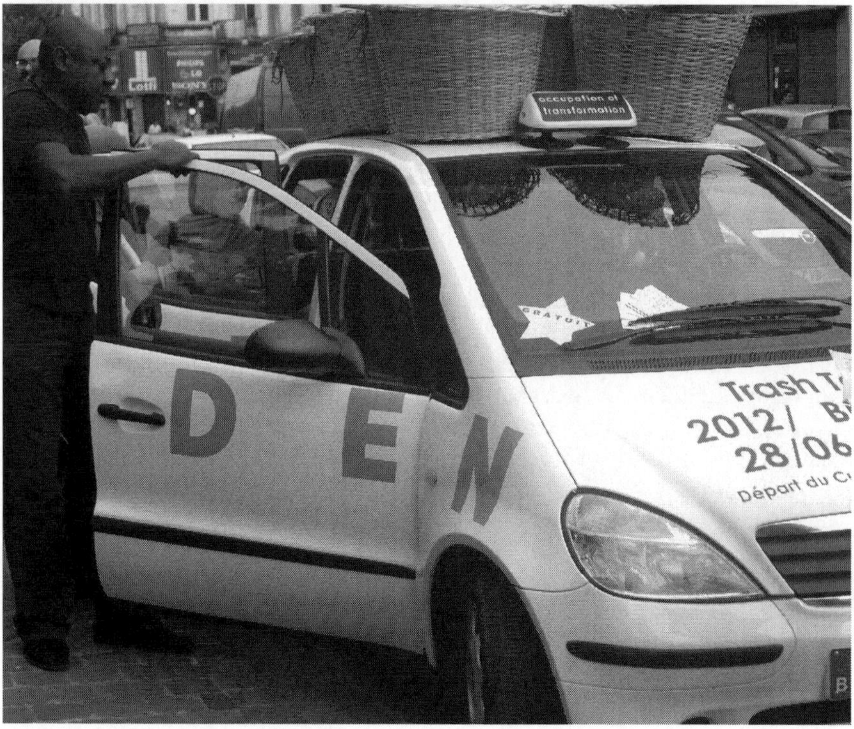

Figure 4.3 Trash taxi, *Wastelanders*, Jeanne van Heeswijk, 2012 (courtesy of Jeanneworks).

The parallels to Benjamin's 'trash aesthetics', as mentioned earlier, here become significant. The particular attention to the marginal and overlooked materiality of urban space was articulated not only through the stories told inside the taxi, but also through the passenger's direct engagement with the trash; in other words, the detritus and backsides of their immediate material and urban surroundings. In line with the artistic practice of Sigman, the trash is here turned into a form of 'social existence' and made to function as conversation pieces to start conversations on the relation between passengers, residents, and the area. As Sigman observed in Brooklyn, the waste in Anderlecht could be seen as an analogy of the residents themselves, as rejected bits and pieces out of place, to be disposed of. Van Heeswijk is critical of the negative effects of upgrading or urban renewal in this regard: 'Often renewal is used to push an agenda of clearance, with some groups no longer being able to take part in that society: an agenda of exclusion' (Viviers 2013). In line with the idea of canonical affordances, the notion of territorial production here points to the way that behaviour and practices that are regarded as either improper or legitimate are supported by material forms and designs. Statements such as 'one does not lie on benches' often imply a tacit specification: 'at this square'. Hence, mobilizing a community is not solely about representing various perspectives and portraying multiple worlds in order to ensure an inclusive practice; it also depends on recognizing the opportunities and limits within our environments for enabling or constraining the creation of such multiple worlds. Accordingly, van Heeswijk wanted the residents of Anderlecht to become 'stewards of public space' (Stanhope 2015), stating that the city belongs to those who inhabit it.

Conclusions

In this chapter, I have scrutinized different collaborative approaches that socially engaged art may apply in order to engage publics in processes of meaning-making in relation to their everyday surroundings. As the challenges faced by Group Material and Ai illustrate, collaborative artistic work requires careful negotiation of the politics of representation, in terms of how to mobilize and represent communities. Lefebvre's notion of a socially lived space here points to the importance for artistic practice, as well as the collaborative process, to engage with the conceptual realms of art and urban space, as well as the material activities within them. The performativity of production is central here, in order to recognize the opportunities as well as the constraints available within production. The practice of Ai illustrates the importance of critically questioning what kinds of subject-positions are made available for participants in socially engaged art in relation to whether the collaboration enables participants to take part in the conceptual as well as in the material production of the artwork. Van Heeswijk's work further emphasizes that the collective narratives produced through socially engaged art are, and need to be, closely linked to the material realities of a particular area.

The discussions in this chapter, then, make explicit the subtle relations of power and representation that exist between artists and their non-artist collaborators in terms of engaging publics in artistic practice. In the next part of this book, I will turn to my own practice with zURBS in order to further scrutinize the participatory processes produced by and within socially engaged art.

Bibliography

Anderson, B. 1991. *Imagined Communities*, 2nd edition. London: Verso.

Architecture Workroom Brussels. 2012. *Garden Parckdesign 2012*. Available at: www.architectureworkroom.eu/en/work/g_a_r_d_e_n_parckdesign_2012/ [accessed 10.06.2016].

Ashford, D. 2010. "An Artwork Is a Person." In *Show and Tell: A Chronicle of Group Material*, edited by J. Ault, 220–225. London: Four Corners Books.

——— and Mohaiemen, N. 2008. *Collectives in Atomised Time*. Ramon Parramon: Ormograf.

Ault, J. 2010. *Show and Tell: A Chronicle of Group Material*. London: Four Corners Books.

Avgikos, J. 1995. "Group Material Timeline: Activism as a Work of Art." In *But Is It Art? The Spirit of Art as Activism*, edited by N. Felshin, 85–116. Seattle, WA: Bay Press.

Bishop, C. 2006. "The Social Turn: Collaboration and Its Discontent." *Artforum* 44 (6): 178–183.

Bo, Z. 2009. "Creating Publicness: From the Stars Event to Recent Socially Engaged Art." 9 (5): 71–85.

———. 2012. "From Gongren to Gongmin: A Comparative Analysis of Ai Weiwei's Sunflower Seeds and Nian." *Journal of Visual Art Practice* 11 (2): 117–133.

Bonnett, A. 1992. "Art, Ideology, and Everyday Space: Subversive Tendencies from Dada to Postmodernism." *Environment and Planning D: Society and Space* 10: 69–86.

Bourriaud, N. 2002. *Relational Aesthetics*. Dijon: Les Presses du Réel.

Bowler, A. and McBurney, B. 1991. "Gentrification and the Avant-Garde in New York's East Village: The Good, the Bad and the Ugly." *Theory, Culture & Society* 8: 49–77.

Callahan, W. A. 2014. "Citizen Ai: Warrior, Jester, and Middleman." *Journal of Asian Studies* 73 (4): 899–920.

Clifford, J. 1981. "On Ethnographic Surrealism." *Society for Comparative Studies in Society and History* 23 (4): 539–564.

Dean, J. 2010. *Blog Theory*. Cambridge: Polity Press.

Deutsche, R. 1988. "Uneven Development: Public Art in New York City." *October* 47: 3–52.

Elden, S. 2004. *Understanding Henri Lefebvre*. London: Continuum.

Gandolfi, E. 2010. "A Dialogue between Emiliano Gandolfi and Jeanne van Heeswijk." In *Visible: Where Art Leaves Its Own Field and Becomes Visible as Part of Something Else*, edited by A. Burtscher and J. Wielander, 133–139. Berlin: Sternberg Press.

Green, A. 2011. "Citizen Artists: Group Material." *Afterall: A Journal of Art, Context and Enquiry* 26: 17–25.

Hertmans, S. 2011. "The Critique of Commitment as Gesture." In *Art and Activism in the Age of Globalization*, edited by L. De Cauter, R. De Roo, and K. Vanhaesebrouck, 54–63. Rotterdam: NAi Publishers.

Highmore, B. 2002. *Everyday Life and Cultural Theory*. London: Routledge.

Joseph, M. 2002. *Against the Romance of Community*. Minneapolis, MN: University of Minnesota Press.

Kanngieser, A. 2012. "A Sonic Geography of Voice: Towards an Affective Politics." *Progress in Human Geography* 36 (3): 336–353.

Kärrholm, M. 2007. "The Materiality of Territorial Production: A Conceptual Discussion of Territoriality, Materiality, and the Everyday Life of Public Space." *Space and Culture* 10 (4): 437–453.

Kester, G. 1999. "Dialogical Aesthetics: A Critical Framework for Littoral Art." *Variant 9 Supplement* 2 (9): 82–123.

Kwon, M. 1997. "One Place after Another: Notes on Site Specificity." *October* 80: 85–110.

LaBelle, B. 2010. *Acoustic Politics: Sound Culture and Everyday Life*. New York: Continuum.

Lippard, L. R. and Chandler, J. 1968. "The Dematerialization of Art." *Art International* 12 (2): 31–36.

Rasmussen, M. B. 2004. "The Situationist International, Surrealism, and the Difficult Fusion of Art and Politics." *Oxford Art Journal* 27 (3): 365–387.

Read, A. 1993. *Theatre and Everyday Life: An Ethics of Performance*. London: Routledge.

Rollins, T. 1982. "Caution! Alternative Space!" In *ABC No Rio Dinero: The Story of a Lower East Side Art Gallery*, edited by A. Moore and M. Miller, 22. New York: ABC No Rio with Collaborative Projects.

Rollins, T. 2010. "What Was to Be Done?" In *Show and Tell: A Chronicle of Group Material*, edited by J. Ault, 217–219. London: Four Corners Books.

Stanhope, Z. 2015. "Questions for Engaging Publics: An Interview with Jeanne van Heeswijk." In *Engaging Publics. Public Engagement*, edited by Z. Stanhope, 11–20. Auckland: Auckland Art Gallery.

Strafella, G. and Berg, D. 2015. "'Twitter Bodhisattva': Ai Weiwei's Media Politics." *Asian Studies Review* 39 (1): 138–157.

van Heeswijk, J. 2011. *Art and Social Change: Learning Collectively to Take Responsibility*. Available at: www.jeanneworks.net/#/essays/art_and_social_change:_learning_collectively_to_take_responsibility/ [accessed 10.06.2016].

———. 2012. "The Artist Will Have to Chose Whom to Serve." In *Social Housing – Housing the Social: Art, Property and Spatial Justice*, edited by A. Phillips and F. Erdemci, 78–89. Berlin: Sternberg Press/SKOR.

Viviers, A. 2013. *Stop Waiting, Start Making: Lessons in Liveability from Jeanne van Heeswijk*. Available at: www.jeanneworks.net/essays/stop_waiting_start_making%3A_lessons_in_liveability_from_jeanne_van_heeswijk/ [accessed 10.06.2016].

Wang, M. 2017. "The Socially Engaged Practices of Artists in Contemporary China." *Journal of Visual Art Practice* 16 (1): 15–38.

Warner, M. 2002. "Publics and Counterpublics." *Quarterly Journal of Speech* 88 (4): 413–425.

Part III

Participatory processes

Critically engaging with urban space in socially engaged art

(Re)inventing urban democratic practices

In Part II, I asked how socially engaged art can challenge the dominant singular and unitary territorial production of urban space by enabling users and residents to recognize the ways in which the city can be seen as a social form rather than as an organization of natural and preordained objects. I addressed this question by focusing on processes of meaning-making. I pointed to the complexity faced by artistic practices that at the same time try to give meaning to the material world and involve individuals in the process. I argued that participation is not solely about portraying multiple worlds and various perspectives, but that artistic practice also has to recognize how the material environment enables or constrains the creation of these multiple worlds.

In this part, I will discuss the challenges posed by this take on participation in relation to the participatory processes in my own artistic practice. This discussion centres on my second set of research questions, which focus on the participatory process of critically engaging with urban space. The notion of territorial production here opens up for examination the multiple legal, physical, or symbolic means through which urban space permits access by certain social groups for selected purposes while excluding others. There is a tendency for groups to be impelled towards like-with-like association, leading to strategies of disaffiliation and colonization in the city (e.g. see Atkinson 2006; Deutsche 1988, 1996; Harvey 1996, 2000). This form of segregation is not solely oriented around the concentration of urban poor or particular ethnic groups, but also the choices of the affluent to withdraw into increasingly insulated enclaves. According to Rowland Atkinson (2006), neoliberal urban policy supports these residential preferences for separation in order to encourage high-income groups to settle in city centres. The need to develop 'lively cities' with a distinctive 'buzz' to them, promoting high-quality design, central-city reclamation of land, revalorization of architectural heritage, and environmental improvements, is directly linked to the attempt to make a particular social group invest financially and socially in the life of the city centre. These initiatives create an image of new urban sites that suppresses their conflictual character. The use of urban space is here restricted to those who are considered its natural owners, denying access for people who are considered 'out of place' (Cresswell 1996), and thus can be constructed as products or bringers of conflict. How can socially engaged art challenge these restrictions by facilitating a participatory process in which a broad variety of residents can articulate and negotiate their attachments to and uses of urban space? I will address this question through practical examples from the workshops carried out as part of the project *invisible Zürichs* that I developed and curated with zURBS.

invisible Zürichs

invisible Zürichs was a two-month-long artistic programme that was part of a residency for independent artistic collectives at the theatre Gessnerallee in Zurich in the autumn of 2013. The theatre provided zURBS with a small but significant budget, as well as one of the main theatre stages for two months, in order to create an urban laboratory that would engage the residents of Zurich in rethinking and rediscovering their city. While *invisible Zürichs* consisted of a whole range of events and contributors, I consider the weekly workshops (14 in total) that zURBS organized in different neighbourhoods in Zurich to be the core of the project. These workshops aimed to facilitate a participatory process in which the participants were enabled to directly engage with the complex social and material structure of their everyday urban environment. The dynamics between processes of meaning-making and the engagement of the workshop participants with urban space here complement the previous chapter's discussion on how artistic practice can help us understand and challenge collective meanings relating to our material surroundings. This perspective foregrounds the political potential of socially engaged art, in terms of how projects such as *invisible Zürichs* may initiate a participatory process in which the participants are enabled to (re)configure a specific space, (re)frame a particular sphere of experience, and (re)think what is taken for granted and designated as a 'natural' condition that we cannot influence.

The political potential of *invisible Zürichs* can be seen in relation to the possibilities and limits the participants have to 'take hold' of, to participate in and to appropriate their urban environment. Lefebvre's notion of a socially lived space is here linked with the concept of urban democracy. In line with the discussions of democracy in Part I of this book, I understand urban democracy as a constant process in which people move towards a better way of living together. As Purcell (2006) points out, there is nothing inherently important about the urban scale in creating a more democratic society. However, as previously discussed, the neoliberal agenda of market-led and technocratic solutions to governance and development has been pursued particularly vehemently in cities. The aim of producing harmonious cities results in urban space being seen as an organic unity, denying the legitimacy of spatial contests. Hence, there is a particular need to (re)invent democratic practices in cities. This part, then, examines how socially engaged art may contribute to urban democracy through a participatory process that has the potential to produce heterogeneous urban geographies based on a variegated politics of identity and difference.

Zurich as a natural urban order

As the International Network for Urban Research and Action (INURA 2008) points out, Zurich has, in recent decades, pursued a politics of growth, struggling to maintain or improve its role as a global city. As part of its striving for international prestige and recognition, the city government of Zurich promotes an image of a clean and orderly city, promoting an urban realm of new, clean, and glossy commodities. The main downtown street, Bahnhofstrasse, happens to be one of the world's most expensive and exclusive shopping avenues. Along with most of downtown, Bahnhofstrasse is constituted by a host of designed spaces, such as shopping centres and heritage sites, that seem to produce familiar and homogeneous sensual experiences. Harsh and unpleasant sensations are kept at bay by the production of moderated urban scenes. In 2012, Zurich was ranked first on Monocle's 'Quality of Life Survey', after featuring in several surveys naming it the city with the best quality of life in the world. However, as Deutsche (1996) points out, the slogan 'quality of life' 'embodies a profound antipathy to rights and pluralism':

> Formulated in the singular, 'the quality of life' assumes a universal city dweller who is equated with 'the public' – identities that the phrase actually invents. The universality of this urban resident is called into question when we note that those who champion a better quality of life do not defend all public institutions equally. While conservative journalists routinely seek to protect municipal parks, they do not necessarily support public education, for example, or public housing.
>
> (p. 276)

As mentioned previously, Deutsche is wary of how urban public spaces are endowed with substantive sources of unity. Certain uses of space are deemed self-evident and uniformly beneficial because they are said to be based on some absolute foundation, such as the 'public good'. INURA observes this approach in Zurich: large parts of the inner-city areas are becoming privileged spaces for a well-to-do urban middle class, following a strategy of 'upgrading distressed neighbourhoods' implemented by the city government as an attempt to strengthen the social fabric by luring 'stable' and affluent residents. As the census data from the city of Zurich (Zürich 2010) suggest, the areas with the highest concentration of immigrants are located in the city-districts to the west and north, outside of the city centre. These are the areas with the highest population density and are also considered to have the highest criminal offence rates. The inner city, on the other hand, is one of the areas with the lowest concentration of immigrants, together with the 'Zurich Gold Coast', which is known for being one of the most affluent areas of Zurich.

Having grown up in Zurich, Sabeth had on several occasions felt suffocated by the perceived conservatism and rigid intrinsic uses of the city. A substantial part of her critique was its lack of 'free space' ('Freiraum'), pointing to the rigid rules of conduct in public space. Through our work with zURBS and with other artistic practices carrying out interventions in urban space, Nina and I could easily identify with her frustrations. Gaining permission for any kind of artistic intervention in public space is a tedious and expensive process, while doing interventions in public space without permission is nearly impossible. The intervention *zuRiche* by the French artist and activist group boijeot.renauld.turon perfectly illustrates this. I had invited the group to Zurich as part of the conference reART:theURBAN (2012) that I was co-organizing. The intervention was based on slow movement through urban space with a set of wooden furniture (chairs, benches, tables, a bed). Passers-by were invited to use the furniture as they pleased and at the same time engage in conversations with the artists. Due to the constant (albeit slow) movement of the furniture through urban space, the artists did not apply for permission for the intervention. After all, *theoretically*, they were not occupying public space, but simply moving through it. The police interfered after 37 minutes, demanding that the artists stop their action. This was the first time the artists (having previously done the intervention in, among others, Venice, Brussels, and Nancy) had encountered such quick and substantial resistance.

Based on this strict control of urban space in Zurich, we – the zURBS team – asked ourselves whether the restrictions we experienced influenced how the residents live in and think about their city. If so, how could we enable the residents to act on other possibilities for being? And how could we make more explicit the ways that different forms of power relations, norms, and mechanisms of control shape how the residents live in and think about the city?

In order to elucidate these questions, we wanted the *invisible Zürichs* workshops to question taken-for-granted assumptions about how we behave, act in, and use urban space in Zurich. The workshops were to facilitate shifts in identification that would emerge, as previously mentioned, through affective registers that would free embodied practices from their sedimented patterns and create a 'micro politics of self-transformation' (Cameron and Gibson 2005, 320). However, we were weary of having the workshops create the kinds of 'isolated islands of empowerment' as critiqued in Chapter 1. The question for zURBS, then, was how to create a socially engaged art that would encourage the participants to directly engage with the complex social and material structure of their everyday urban environment in order to recognize, question, and challenge the material and social circumstances that regulate their everyday actions and behaviours, and to do so in ways that weren't temporally or spatially isolated.

Layering complexity

There are several layers of complexity at play in the following two chapters. First, the complexity faced by zURBS as we tried to develop an open participatory and artistic process that would engage the residents of Zurich in questioning their urban environment and potentially subverting their perception of the dominant urban sensory order or pointing to alternate orders. Second, the complexity of trying to align my practical and often frustrated, contradictory, intuitive, and incoherent reflections and thoughts about our practice with theoretical arguments and perspectives. Third, the complexity of trying to observe, analyse, understand, and communicate the participants' experiences of the workshops, in terms of narratives told during the workshop and post-workshop interviews, as well as narratives that were not told, but expressed or mediated through the found objects, drawings, photographs, scribbled stories, and comments. Finally, the fourth layer of complexity relates to the question of how to communicate all these different layers of complexity so that they can be read and understood by the readers of this book. To communicate these layers of complexity, this part of the book bears several resemblances to the 'alternative city archive', which I will call 'the archive' for brevity, that was constructed through the workshops in *invisible Zürichs*. The next two chapters express the simultaneous coexistence of many different and possible experiences, perspectives, and stories (both told and untold). Accordingly, it consists of loose ends, missing links, and uneven, conflicting, unassimilable, but related perspectives. These perspectives are not meant to add up to one homogeneous story, critique, or take on the workshops. Rather, they display how practice as research requires the researcher to talk of 'might' and 'if', of mess and what is missing, of tensions and contradictions, of gaps and bridges between different worlds, of stories lost and stories retold, of slippage and fluidity. As with the workshops, this is an experiment in meaning-making in which the 'outcome' is necessarily unpredictable. Visitors to such environments must play an active role as navigators, wayfinders, and meaning-makers, drawing their own observations and conclusions without the reassuring presence of an authority to which to defer. This demands a form of explorative engagement that is oriented towards the experience of unusual and new features, in order to open up new understandings and perspectives.

Bibliography

Atkinson, R. 2006. "Padding the Bunker: Strategies of Middle-Class Disaffiliation and Colonisation in the City." *Urban Studies* 43: 819–832.

Cameron, J. and Gibson, K. 2005. "Participatory Action Research in a Poststructuralist Vein." *Geoforum* 36 (3): 315–331.

Cresswell, T. 1996. *In Place. Out of Place*. Minneapolis, MN: University of Minnesota Press.

Deutsche, R. 1988. "Uneven Development: Public Art in New York City." *October* 47: 3–52.

———. 1996. *Evictions: Art and Spatial Politics*. Cambridge, MA: MIT Press.

Harvey, D. 1996. *Justice, Nature and the Geography of Difference*. Oxford: Blackwell.

———. 2000. *Spaces of Hope*. Edinburgh: Edinburgh University Press.

INURA. 2008. *The New Metropolitan Mainstream*. Available at: www.inura.org/NMM_Posters_PDF/INURA11_Zurich.pdf [accessed 09.10.2018].

Purcell, M. 2006. "Urban Democracy and the Local Trap." *Urban Studies* 43 (11): 1921–1941.

Zürich. 2010. *Statistiches Jahrbuch Zürich 2010*. Zurich: Statistik Stadt.

5 Contesting the city as a natural urban order

Arriving and orientating

You have entered the 'alternative city archive'. The location – the south stage of the theatre Gessnerallee – is familiar to you as you have seen plays put up here before. But now the stage has changed into a messy room filled with 'stuff'. The walls are plastered with photographs, notes, and drawings. An ironing board, a bag full of plastic bottles, a street magazine, a hubcap of a car wheel, and various cardboard boxes are hanging from the ceiling. Small selections of garbage, mainly tin cans and chips bags, are organized in little groups fenced in with gaffer tape on the floor. As your eyes scan the room, you notice more and more details: the small glass jars with various things in them, the worn-out shoes, the wine glass with some red wine still left in it ...

> I must say that in the beginning when I entered the archive and was told to look around I felt a little bit lost [. . .] But then I noticed more and more little details and it became like an adventure, it was just that I had to introduce myself to the stuff and the stuff had to introduce itself to me . . . I even took some pictures of the archive because I found so much funny stuff there, like for example that drawing with the unicorn.
>
> (Hannah, 30s, HR worker, personal interview)

The atmosphere is hectic; the workshop facilitators, dressed in white painting overalls and calling themselves 'researchers of invisibilities and imaginations', are busy archiving. You are asked to fill in a 'researcher profile'. 'OK, so I am a researcher now', you think, as you curiously glance at the strangers that are your fellow researchers.

One facilitator gathers everyone's attention. She tells you about the book Invisible Cities *by Italo Calvino, in which the narrator, Marco Polo, describes various imaginary cities that are in fact stories of the very same city, his hometown Venice. 'Our thesis is that Zurich consists of even more invisible cities based on the many different perspectives, realities, and imaginaries of its inhabitants! Now your task is to go out there and find traces of them. These are traces*

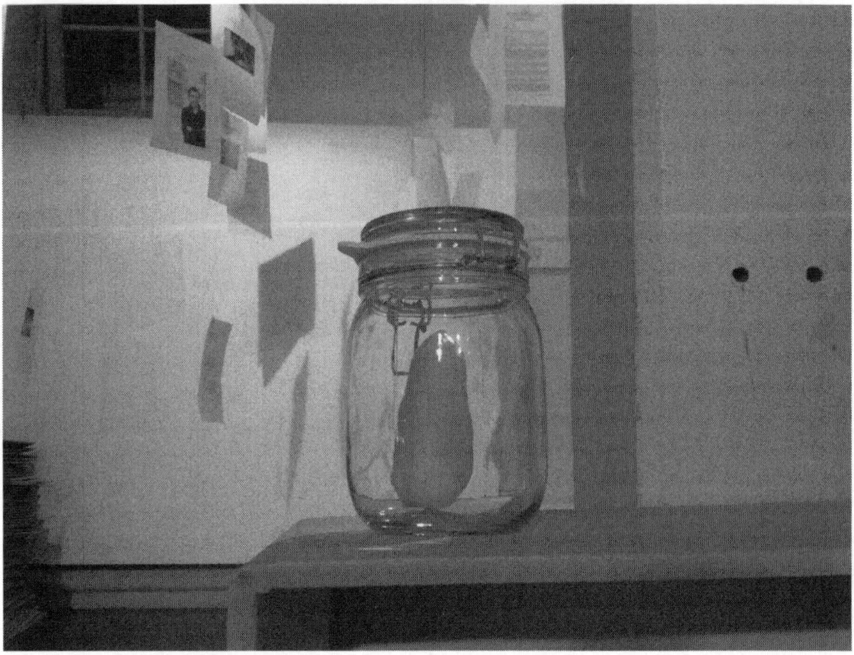

Figure 5.1 Tomato in a jar, detail, 'the alternative city archive', *invisible Zürichs*, zURBS, 2013 (image author's own).

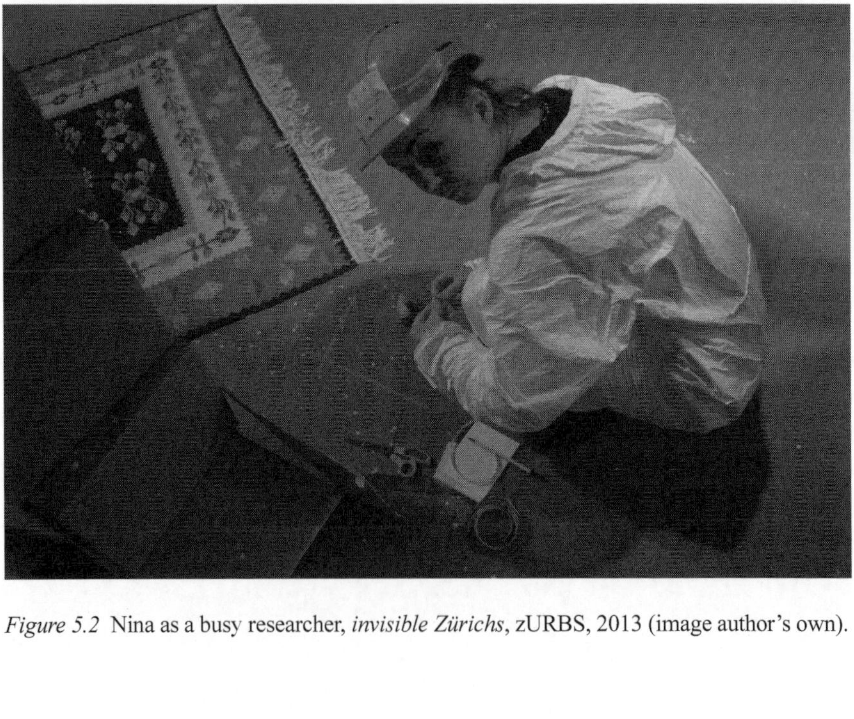

Figure 5.2 Nina as a busy researcher, *invisible Zürichs*, zURBS, 2013 (image author's own).

of hidden lives, forgotten memories, eaten food, future dreams, weird sounds, uncanny atmospheres, everyday stories, and so on.'

The facilitator shows you examples from 'the archive' and what previous researchers have collected: 'Look at this collage containing a pea and a peanut placed in a cartoon that tells the story of how they met and fell in love – one of the many love affairs of invisible Zürichs*!'*

> What I got was that it was a research also in your own mind, about how you interpret the city and how it speaks to you. And it came through examples, like the one you showed with the peanut [. . .] So this was for me, OK, OK, so it must be about my own creative, poetic interpretation of the city what we were going to embark on. And also, the way things were presented already brought me a lot into the details of the city, rather than the big meta structures.
>
> (Ane, 40s, artist, personal interview)

You are randomly divided into groups of three to five people and handed a tote bag containing a map of a certain area of Zurich. The map, you are told, indicates the different places that your group could go to in order to search for the invisibilities. 'But how to orientate oneself with this map?' you wonder. It has got no place names on it, and it is hard to distinguish one street from the other

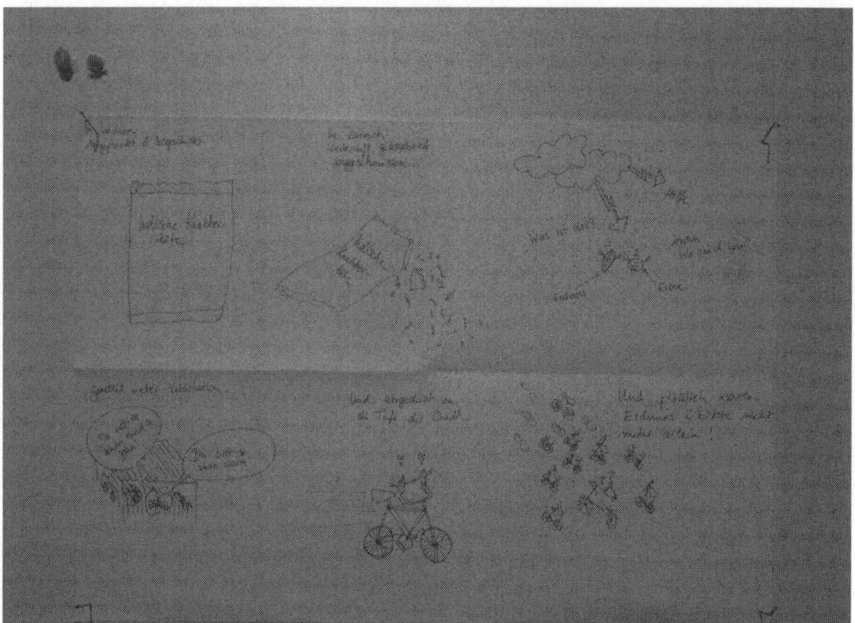

Figure 5.3 Peanut story made by participants, detail, 'the alternative city archive', *invisible Zürichs*, zURBS, 2013 (image author's own).

as they are simply indicated with lines that remove any hierarchy in terms of main streets and side streets.

> What I truly loved with this experience was that it was open, in terms of interpretation [. . .] I loved the fact that the map was not clear.
>
> (Eirini, 30s, game designer, personal interview)

Your group enters the city, somewhat hesitantly. You have jointly decided that you will try to find the different places marked on the map as you have been informed that there are envelopes hidden here. The envelopes contain inputs that might be useful for your search. As you are walking through the streets, trying to orientate yourself in a strange map and looking for the hidden envelopes, you realize that you feel like a stranger in your own city.

> For me it was a new way to see my own city where I lived for years and years – more or less 50 years that I have been living in Zurich. I just took the chance to look at it from another point [of view] . . . If you go through the city with a certain concept that you prepared for us, you step into a place that you normally wouldn't have stepped in if you were just doing your business in Zurich, and this was just amazing to see. Let's say – for instance – this parking house where I passed I do not know how many thousand times, and that I have never been into. Or this hostel . . . To have a certain view on Zurich, this touristic view . . . which I never had because I am living here. So that was . . . I found this very funny.
>
> (Jürg, 60s, social researcher, personal interview)

> This makes me think about this process of travelling. If you consider this project as a social practice, it is more or less the same idea as tourism as a social practice. When we are tourists we also explore the urban space in a new perspective because we are going elsewhere . . . Where tourism comes from is to see space from a new way, to explore something new and being adventurous. And this actually means that you can be a tourist in your own city just by going around looking at it differently and changing your perspectives and your categories . . . You get multiple perspectives on the same space, so you always reimagine the same space.
>
> (Sara, 20s, student, personal interview)

Rearticulating the everyday

The workshops provided a context for reflecting upon, discussing, and experiencing Zurich that adopted imaginative, transgressive, and poetic constructs. The aim was to build alternative and personal positions from which to experience a place so as to question its 'taken-for-grantedness'. Several participants mentioned that the workshops made them feel like tourists in their own city in this regard. The idea of tourism invoked by the participants here relates to seeing the

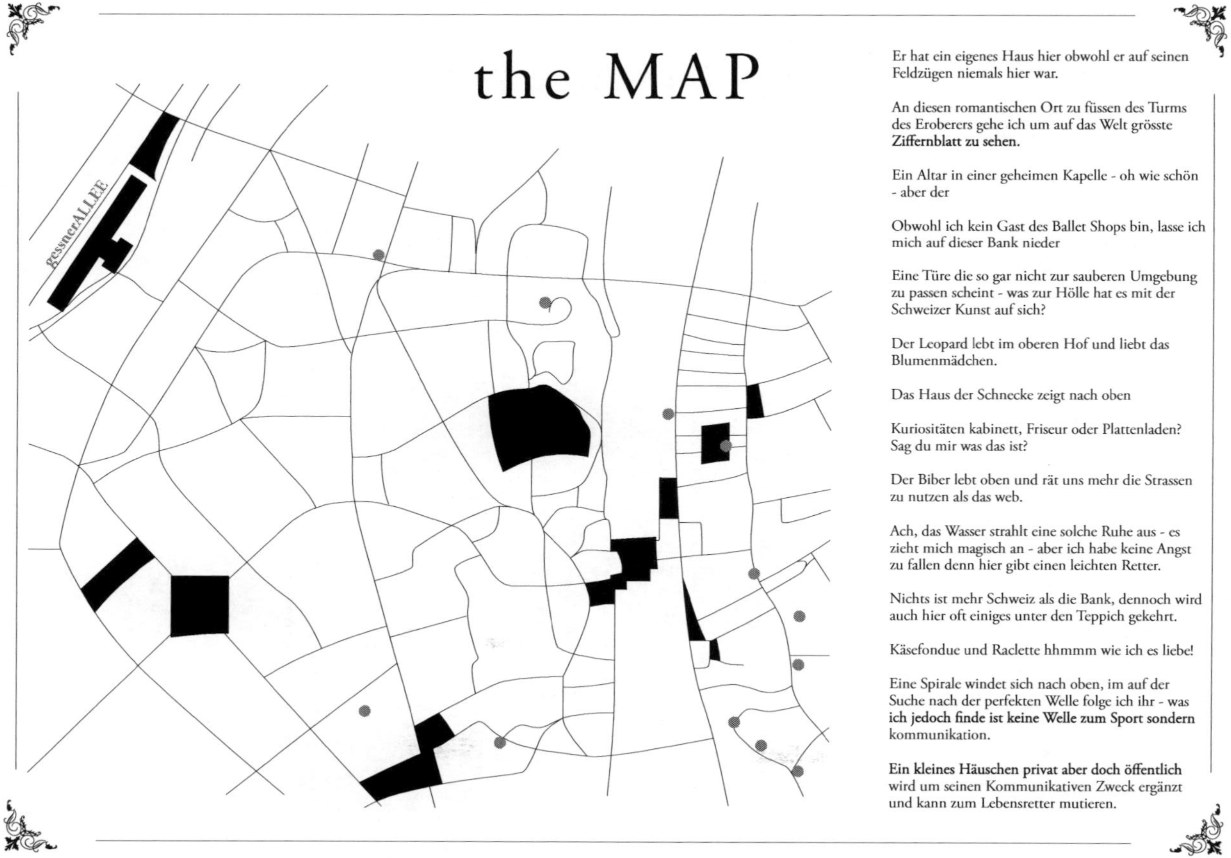

Figure 5.4 The map, *invisible Zürichs*, zURBS, 2013 (image author's own).

city from a new perspective, to discover, explore, and to be adventurous. This understanding of tourism is quite different from how tourism is often discussed in academic and popular discourse. In these discourses, tourism is often seen as fashioned by culturally coded and commodified escape attempts, which involve unreflexive, habitual, and practical enactions, reflecting common-sense understandings of how to be a tourist (Edensor 2001). The sociologist John Urry (2002) points to how being a tourist involves the notion of 'departure', 'of a limited breaking with established routines and practices of everyday life and allowing one's senses to engage with a set of stimuli that contrasts with the everyday and the mundane' (p. 2). However, when tourists enter particular locations, they are usually informed by pre-existing discursive, practical, embodied norms and a series of recommendations (e.g. travel guides) that guide their orientations and achieve a consensus about what to do.

This removal from the quotidian through prescribed experiences is quite different from the defamiliarization of the urban everyday that we wanted to cultivate through our workshops. Rather than facilitating an escape from the everyday, our workshops aimed to rearticulate and reconceptualize what is familiar to us. By providing the participants with an abstract map of Zurich, we attempted to counter the 'tourist guide machinery' of discursive, regulatory, and practical norms, which attempts to direct the experience of a place. Rather than giving a series of recommendations, suggesting that the participants behave in particular ways, the map made the familiar city of Zurich strange. Like the book *Invisible Cities*, the map opened up multiple perspectives and ways to get to know and reimagine the city. To privilege this plurality is not to leave the real city behind, but rather to insist that our experiences of cities are caught in dense metaphorical meanings and the accumulation of images and signs (Highmore 2005).

The tourist experience the participants referred to, then, can be seen in the light of this plural form of explorative engagement with urban space. By focusing on multiple perspectives, the workshops implicitly critiqued the highly regulated and controlled urban environment in Zurich that does away with the city's surprises, detours, and disturbances. Richard Wentworth reflects on this feeling of control that is imbued in the contemporary day-to-day activities of Zurich:

> I actually feel – mentally – slightly ill when I am in Zurich, and I am thinking there must be some bad things going on there because it cannot be that nice. I think there is this little policeman inside everyone's head; like, when I go on the tram and I see in all these people that I am not getting on the tram quite right. I don't know what it is, but I see judgementalness. And I grew up in a very judgemental household, so I am very quick to detect it once it is there. And how to deal with the ticket, should I validate it?
>
> (personal interview, 21.10.2014)

Wentworth's observation points to how urban space feels oppressive and humiliating as it is perceived to have a single meaning that necessitates proper uses or proper places for its users. The French sociologist Laurent Thévenot (2014) points to a tension between familiar formats and personal engagements with our surroundings and the more communal engagements that rely on public commitment. Whereas the former relies on convenient appropriation, the latter demands a standardized and 'proper' use according to a public judgement of what is right or wrong to do at this particular place. Personal views and engagements are here sacrificed for a characterization of the common and public good – the existing spatial structure has to be 'beneficial to all' (Thévenot 2014). However, this idea of a 'proper use' obscures the conflictual manner in which urban space is defined and used, repudiating the existence of personal engagements that counter dominant uses of space. Accordingly, urban space may prove oppressive and humiliating for people who use space in other ways than what is defined as its 'proper' use. A classic example in this regard is how the former mayor of New York, Ed Koch, appealed to 'common sense' when stating that it is only reasonable for the authorities to say that (homeless) people cannot stay at the train station unless they are there for transportation (Cresswell 1996; Deutsche 1988, 1996).

We – the zURBS team – wanted our workshops to question this idea of a 'proper' use of space in order to find ways of maintaining a composite and conflicting community that caters for the articulation of the variety of senses of reality and modes of engagement in our lives and our living in common. How could the workshops facilitate a participatory approach that would create the conditions under which a broad variety of residents could articulate their personal attachments and uses of urban space so that these uses and attachments would emerge as 'visible' and 'salient'?

The question of familiarity and participation

For some of the workshops, we were struggling with getting participants. Consequently, we went into the streets to recruit passers-by. One day, we succeeded in recruiting an elderly Polish couple and guided them towards Südbühne, the workshop location, which was located 100 metres from where we encountered the couple. Just as we were about to enter the theatre stage, the couple seemed to have gotten second thoughts, and as he put his foot on the doorstep, the man changed his mind and said it was not for them. And so, they walked away. This experience made me question whether our workshops were alienating potential participants by not being receptive to familiar formats. The couple might have changed their minds exactly because they did not feel familiar with the setting of the theatre stage. This may be explained by linking familiarity with the notion of 'orientation'. Being able to orient oneself within the 'appropriateness' of a space or situation is a question of familiarity. As Heidegger (1962, cited in Simonsen 2012, 19) points out, 'I necessarily orient myself both in and from my being already alongside a world which is "familiar"'. Although orientation reaches out towards something/somebody different from ourselves, it does rely on a familiar starting point from which we then can extend our reach in terms of what we can have and do. Hence, orientation is about both 'finding our way' and 'feeling at home' (Simonsen 2012, 19–20). Only when one feels at home, is one prone to undertake what Thèvenot (2014) points to as an 'explorative engagement' (p. 15) that, while remaining personal, is oriented towards the experience of unusual and new features.

It was much easier to engage people who were already familiar with participating in artistic and cultural initiatives in this form of exploratory engagement. In our public workshops, taking place every Saturday at 12.00 from 26 October to 30 November, the majority of the participants belonged to the so-called 'cultural crowd' that is particularly interested in these forms of activities. This 'cultural crowd' was comprised of people such as the normal theatre audience of Gessnerallee, as well as curious students, sceptical architects, enthusiastic artists, provocative activists, retired people with time on their hands, young professionals looking for an out-of-the-ordinary experience, and so on. This was a great crowd to work with as a researcher. They were more than willing to be interviewed about their experiences of the workshop and were often very articulate in their reflections, providing quotes that were perfect for illustrating the transformative potential of the workshops. However, I could not avoid the feeling that these participants sometimes were paying me lip service, and also how could zURBS argue that the point of *invisible Zürichs* was to open up for a multiplicity of knowledges and plurality of perspectives on Zurich if most of the participants belonged to a certain segment of the population?

To ensure a broader range of participants, we organized workshops during the week that targeted particular groups that did not necessarily belong to the 'cultural crowd'. These groups were recruited through the method of 'netwalking' – a term zURBS invented that referred to a form of networking involving walking around the neighbourhood, knocking on doors, and this way getting in contact with organizations, institutions, and groups that are located in the specific area in which the workshop would take place. We often had a hard time explaining the workshops to these groups. Often, the reply to our invitation was, 'We know Zurich, we live here'. When discussing the problem of not getting participants with Laura, the volunteer helping to facilitate the workshops, she pointed out that the problem was the question 'Why should I do it?'. It was not sufficient to convince, for example, the Kenyon hairdressers to take part on the abstract basis of 'changing their perception'. The argument should rather be grounded in people's personal relationship to place: 'This is a shitty place, but by actively using it we can collectively change it for the better'. Laura was right. While trying to avoid stigmatizing certain groups or places by defining them as problematic or subjugated, and thus in need of 'liberation', we ended up in the opposite camp: we framed the aim of the workshops within abstract notions that were not immediately associated with people's personal and familiar concerns or situation. As one of the employees in an Asian retail shop angrily replied when I tried to convince her to take part in the workshop, 'I don't have time to do games, can't you see I'm busy working?' Clearly, a change of perception is not top priority for someone who has enough with making do and getting by.

Alas, we had to admit to a certain level of 'exclusiveness' in our workshops. We could not force all kinds of people to participate. All we could do was try to relate the workshop to people's personal concerns so that potential participants would have the opportunity to evaluate whether the workshop would be relevant and interesting for them to take part in or not. Accordingly, we tailored the aim for the workshop to whatever group we were in contact with. For the schools, we would point out the educational aspects of the workshop; for the book club, we would focus on the Italo Calvino link; for the social workers, we would focus on getting to know the streets; and so on. After a whole summer of netwalking and endless attempts to explain the workshops to groups and individuals that would not normally attend these kinds of cultural events, we managed to recruit social workers from two different neighbourhoods in Zurich, employees and users of a daytime centre for alcoholics, a school class of 16–17-year-old boys, and a group of children from an after-school institution.

Transgressing (b)orders

'I found it!' you call out to your group as you detect an envelope under a garbage container belonging to a fancy upmarket housing block on Zurich's Gold Coast. The envelope contains the story of Leonia – a city that is obsessed with newness; every morning, its citizens wake up between fresh sheets, wash themselves with new soap bars, and put on brand-new clothes. Yesterday's leftovers are neatly put on the pavements in plastic bags, ready to be whisked away by the rubbish trucks. The story is accompanied with a task: 'What traces of opulence can you find? Collect them and label them as if they were archaeological findings.' On your way to the next stop, you find yourself climbing the fence of a closed-off restaurant garden in order to collect a selection of empty champagne bottles. An elderly man in your group picks up a cigar box from a park bench, and a young girl is amused by all the broken Christmas decorations she spots along the streets.

You find another envelope containing the story about the chaste city of Chloe, in which strangers imagine a thousand things about one another. You are asked to make two strangers meet and document the meeting however you see fit. Whether you want to make up a story about this meeting or have two strangers meet for real is completely up to you. Your group spots a man with a leather bag sitting reading the newspaper at a café inside a busy market hall. Collectively, you make up a story about how he is sitting there every day, waiting for an adventure.

> He carries a bag with perfume and clothes in it, so that whenever an adventure would occur he would be ready to follow it. Then, this girl comes along [. . .] She sits next to him and has a coffee. Maybe this is the adventure he has been looking for? But no, they do not talk to each other [. . .] It is a potential for an encounter that never happened.
>
> (Roger, 40s, urban planner, participant observation)

You write down the story while observing its main characters from a distance. One of the women in the group decides to go up to the man and give him the story. The man is amused to hear that he is featuring in your fictional construction,

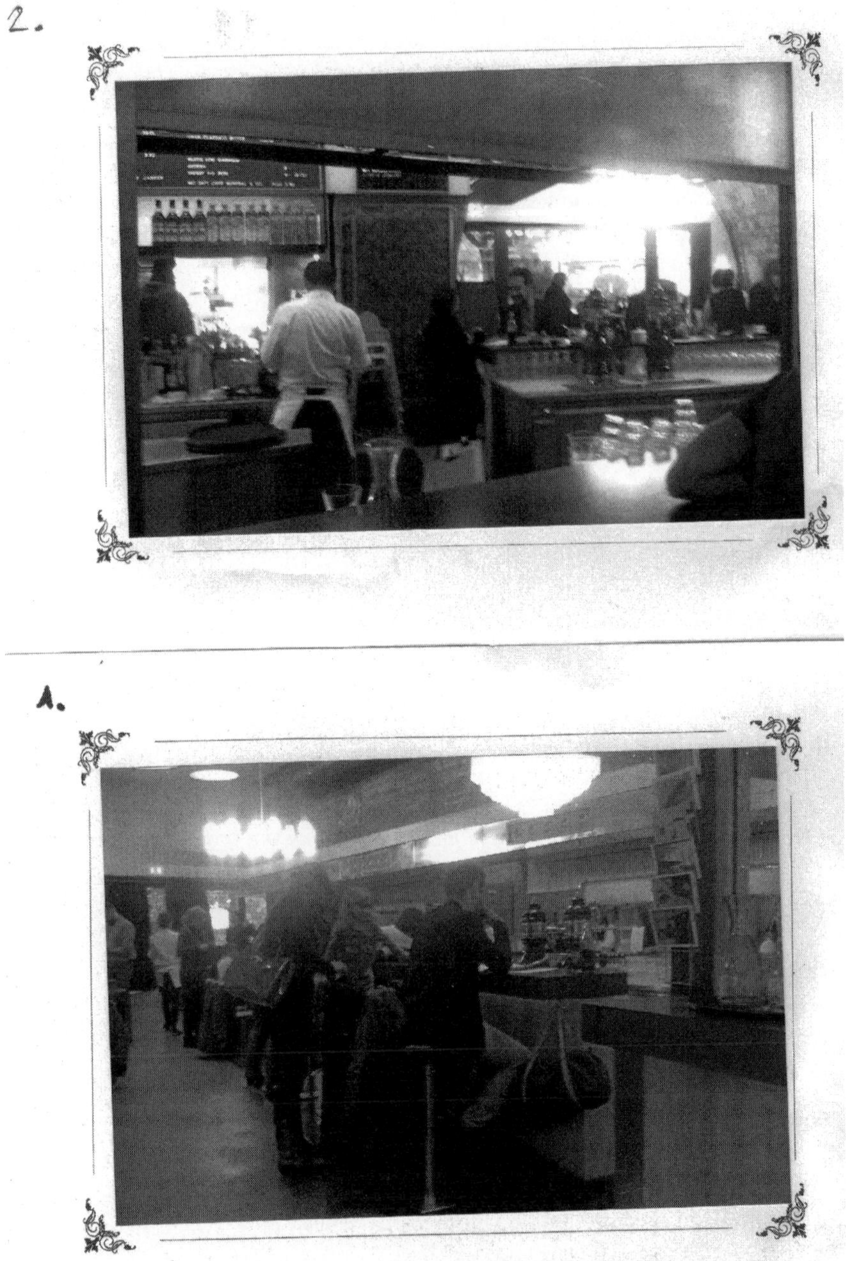

Figure 5.5 Photos by participants of people meeting in a bar, detail, 'the alternative city archive', *invisible Zürichs*, zURBS, 2013 (image author's own).

which is, it turns out, actually pretty close to reality – the man really has got perfume in his bag!

> What really works for me is that you really feel like a scientist or explorer [. . .] This task from you enables me, or us, to do things we wouldn't [otherwise] dare to do. We would never, I would never have this idea to write a story between two persons [. . .] For me, [the stories] were important [. . .] It brings you in a kind of mood which makes maybe fulfilling the task also easier, so it brings the people – at least me – in the right feeling.
>
> (Roger, 40s, urban planner, personal interview)

> The sensibility got more sharp in a way. I was more aware of my surroundings . . . Like, when we were looking for this luxury trash, we found everywhere these plastic glasses, like the ones drunk at parties. And they were left in all kinds of places, like by staircases, next to chairs . . . And I was thinking: why do people leave their party stuff everywhere? It is so irritating and unnecessary!
>
> (Hannah, 30s, HR worker, personal interview)

The hunt for the hidden envelopes encourages you to enter places you wouldn't normally enter, such as a youth hostel for travellers and a local brown run-down pub where the whole clientele turned around and gave you the 'What are you doing here?' *look as you entered. In the search for another envelope, you discovered a crypt underneath the church. Having lived in Zurich for over 20 years, one of the women in your group was amazed to find that such a place was actually publicly accessible.*

> I went into this gallery in order to try to find the envelope. The gallery had a vernissage, but no one noticed me even though I was clearly not looking at art but looking everywhere else for the envelope. I even entered someone's

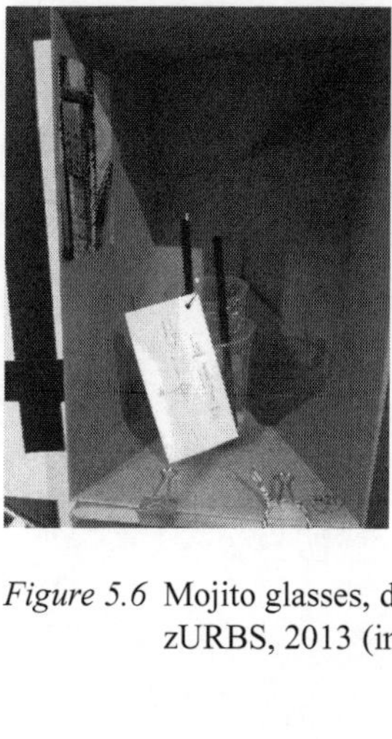

Figure 5.6 Mojito glasses, detail, 'the alternative city archive', *invisible Zürichs*, zURBS, 2013 (image author's own).

office surrounded by office stuff in order to look for the envelope. That was something I would of course *never* have done otherwise, but because I was looking for the envelope I sort of had the 'permission' to do it.

(Josh, 40s, game designer, workshop conversation)

'Governing through standards'

The workshops facilitated several moments when the participants had to transgress social and physical boundaries in order to find the envelopes and carry out the tasks within them. We found these elements of 'transgression' important steps for challenging taken-for-granted assumptions on how we behave in, act in, and use urban space. Transgression is here defined as crossing some line that was not meant to be crossed, and thus making some boundary visible that you might not know existed (Cresswell 1996). Parallels can be drawn to the power of distraction as a mode of cognition that can help people understand and transform their experience, as mentioned in the previous chapter. As Cresswell (1996) points out, our consciousness of place in terms of the 'naturalness' of the environment, defining who and what is in place or out of place, all but disappears when things are working well (p. 10). It is only when things appear to be 'wrong' that we might become aware of the social and normative power structures that regulate the city. The ethnomethodologist Harold Garfinkel here argues that upsetting commonsense assumptions has 'the effect of annoying people and this helps us discover the underlying taken-for-granted assumptions that help mold social action' (cited in Cresswell 1996, 22).

One example in this regard is an experiment we did with zURBS regarding the appropriation of a parking lot for an afternoon. Following the conventional rules of conduct, we paid for the spot by putting money in the parking machine and received a parking receipt. However, we did not park a car on the spot, but used the spot to create an 'asphalt gallery' with empty picture frames. Subsequently we hosted a vernissage on the spot. The intervention caused outrage by car owners looking for a parking place, resulting in one of them – a furious businessman in a Jaguar – calling the police on us. By referring to the operating instructions of the parking lot (it is for cars), our actions were questioned in an attempt to identify 'misuse', and thereby disqualify zURBS as incompetent or out of place. The 'dispute' led the Jaguar driver to make reference to the most legitimate collective conventions, in this case the police and the conventional rules applying to parking lots. This example illustrates how conventional utility may legitimize a singular territorial production of a place. Rather than allowing for several territorial strategies to be produced at the same time and place, there is a sense of one specific territorial production being more correct than others: the parking lot is reserved for cars. By upsetting the balance of common sense in terms of transgressing the predefined use of a parking lot, zURBS put this 'givenness' of the parking lot into question.

Zurich has a strong moral covenant in terms of the 'proper' use of urban space. Thévenot (2014) describes this moral covenant as a form of 'governing through standards' (p. 12), which is based on a generalizing process that relies

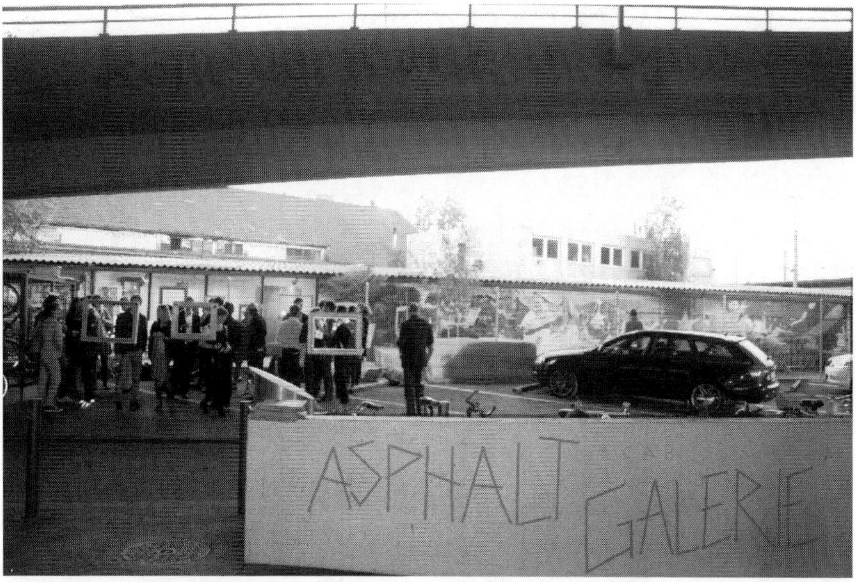

Figure 5.7 The Asphalt Gallery, zURBS, 2012 (image author's own).

on conventional and coordinative forms of equivalence and qualification, such as classifications, standards, and rules. This form of governing depends on formatted and structured environments that support the conventional way of engaging with the surroundings, for example by emphasizing the 'canonical affordances' of objects and things, as mentioned in the previous chapter. Of course, in order to be able to live together with others, it is necessary to standardize information, such as, for example, signs, so that the environment becomes practically legible for all. However, as both Deutsche and Thévenot argue, these procedures of standardization may support official and unofficial mechanisms of control and regulation that instantiate restrictive and oppressive ways of engaging with urban space. The poster campaign 'Erlaubt ist, was nicht stört' ('Permitted is what does not disturb') that from 2000 to 2005 was implemented in Zurich is a good example in this regard.

The campaign was an attempt to maintain a 'clean and orderly' Zurich, occupying urban space with posters promoting the 'Erlaubt ist, was nicht stört' slogan. Some of the posters just depicted a big exclamation mark, suggesting that any kind of 'disturbing' behaviour could be regulated by making reference to the posters. Other posters simply stated 'Ruhe!' ('Quiet!'), 'Ordnung halten in Tram und Bus' ('Keep order in the tram and bus'), 'Abfall in die Kübel' ('Trash in the trash can'), and so on. Some of the posters used a more provocative, albeit humorous, language to prevent people from 'disturbing' public space. For example, those who want to take a pee outdoors were confronted with posters saying

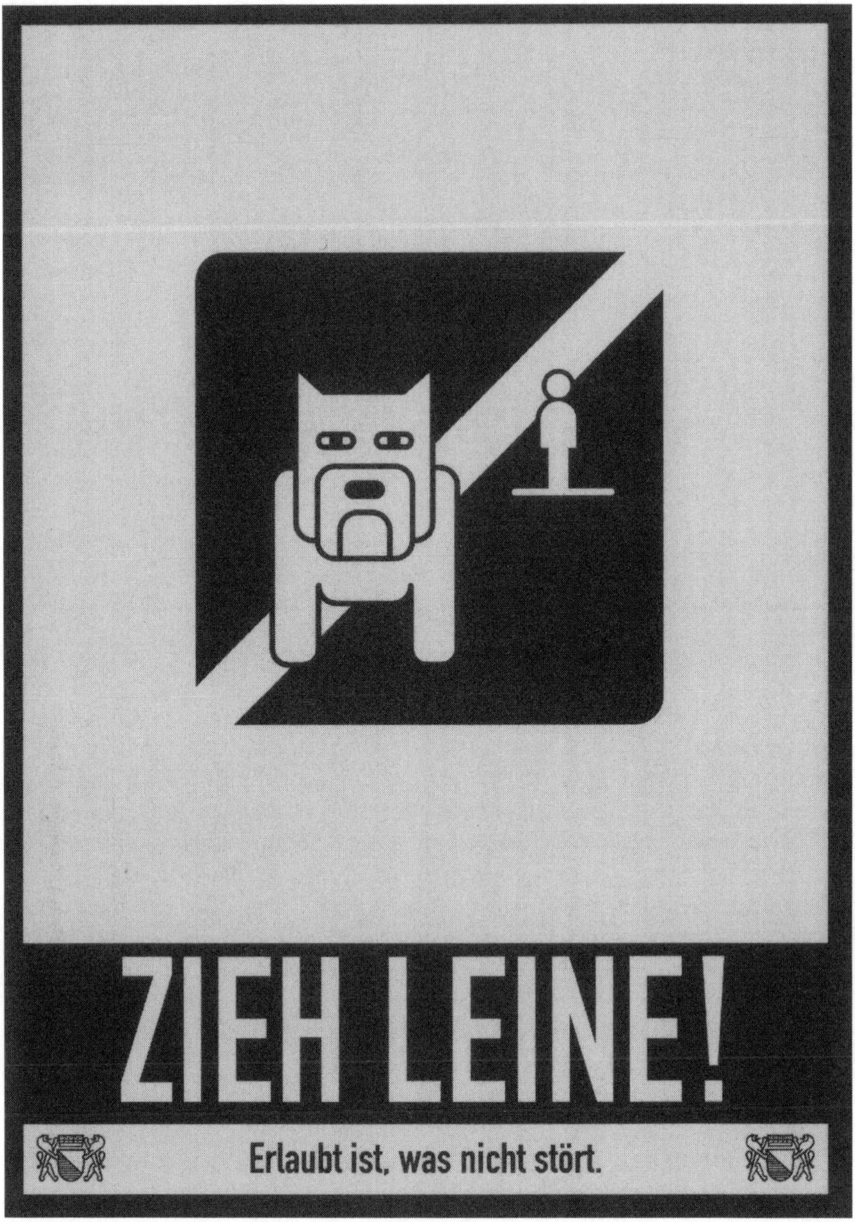

Figure 5.8 *Erlaubtist, was nicht stört*, campaign for safety and cleanliness in public space, poster and post card series by Stadt Zürich, 2000/2001 (© SozArch F 5038-Ka-017).

'Verpiss dich!' ('Piss off!'). Dog owners were faced with a poster saying 'Zieh Leine!' ('Go away!'), which plays on the double meaning of the word 'Leine' as both 'dog leash' and 'away'. In blunt language, the poster thus stated that if your dog is not on a leash, you are not welcome in the area.

In a pamphlet protesting against the campaign, Stadt.labor – a Zurich-based platform for a critical view of urban development – points to how the campaign was used to repudiate those groups who counter dominant uses of space. The campaign, they argue, was a disguise for a zero-tolerance policy in relation to urban fringe groups such as drug users, homeless people, migrants, and – as repeatedly emphasized by the authorities – the punks located in the neighbourhood of Stadelhofen. These punks have a reputation in Zurich for owning several dogs that are not on leashes. The socially constructed meaning of Zurich as clean and orderly here directly affects judgements of the events and people in the city, including what and who are perceived as disturbances to the 'natural order'. The recognition of the punk as a disturbance here works to support the 'proper' residents' fantasy that the city is essentially an organic whole. The punk is constructed as an ideological figure, a negative image created to restore positivity and order to social life.

Campaigns such as 'Erlaubt ist, was nicht stört' risk alienating the city from the social life of its inhabitants by claiming that the city is an independent object that speaks for itself (Deutsche 1988, 6). *invisible Zürichs* aimed to make visible dynamics that challenge this idea, and rather point to how users collectively perform and attach certain 'schemes of perception' to urban space. This was illustrated in a workshop in which one group had found and brought back a plastic bag containing a half-full bag of potato chips. A woman in her 20s started to eat the chips, causing raised eyebrows. The fact that this was a thrown-out bag of potato chips found on the street made it quite unappealing to eat the chips for the majority (I would dare say the rest) of the participants, who did not even touch the content of the bag. Hence, the woman transgressed the boundary of 'normal' behaviour and accordingly challenged the existing meaning of trash found in the street as something 'we' do not want to come into close contact with. Had the chips been bought by the facilitators of the workshop and put in a bowl for the participants to serve themselves, it would have been a different matter. This in spite of the fact that the chips might be even 'dirtier' in terms of bacteria, after having several hands touching them, than the chips in the bag that potentially belonged to one person who bought it as a pre-dinner snack and didn't manage to finish it. These are, of course, speculations, but the episode points to how acts of transgression make visible how places are active forces in defining appropriate practice, and thus also in determining the functional fixity of objects: a bag of potato chips at a party is a consumable object that you would eat, whereas a bag of potato chips found on the street is waste that you should stay away from.

The conception of the workshop participants as a homogeneous whole was also challenged in this situation. The woman eating the chips acted in a way that did not correspond to a behaviour that the rest of the group could identify with. Her actions thus raised questions about difference and 'otherness'.

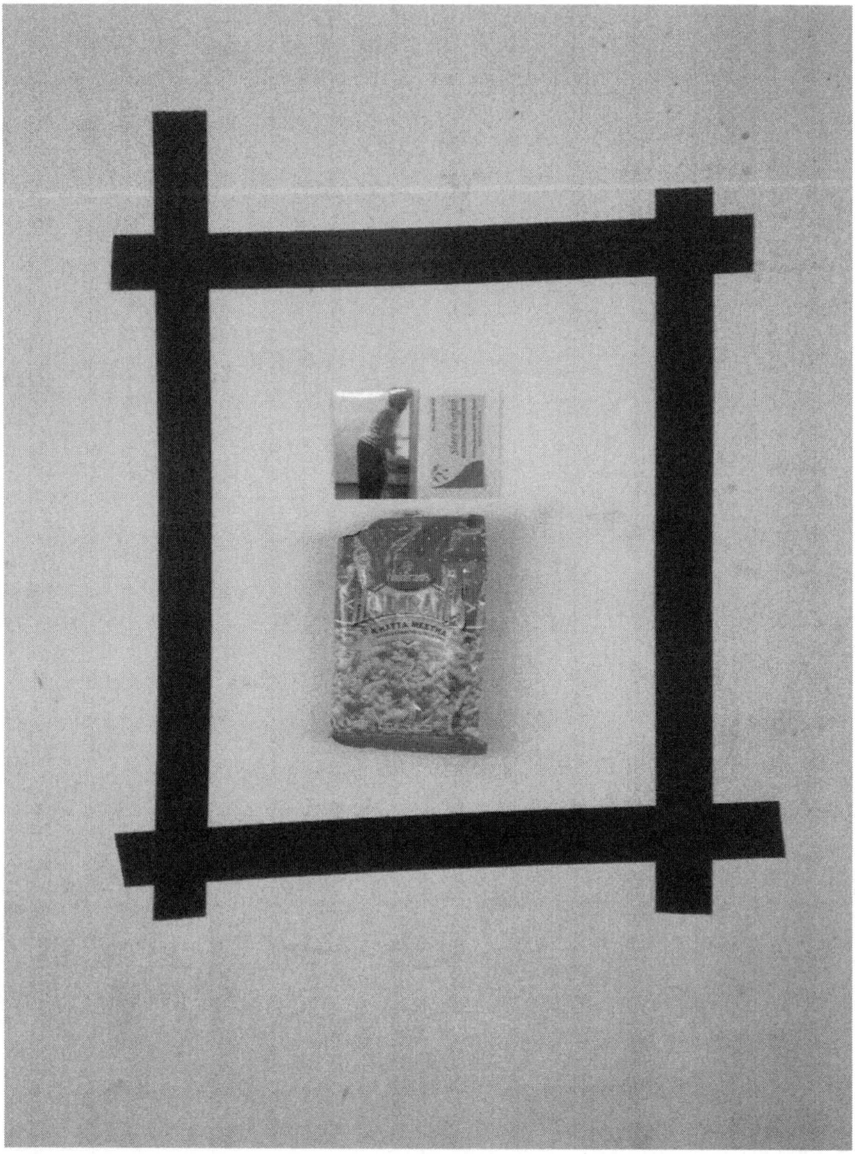

Figure 5.9 Bag of chips, detail, 'the alternative city archive', *invisible Zürichs*, zURBS, 2013 (image author's own).

As Cresswell (1996) observes, conflicts are avoided as long as individuals aspire to that which fits in with what is appropriate for their objective position. However, when subjective beliefs and actions do not correspond to the objective position, a conflict is introduced that questions the 'naturalness' of certain behaviours and the 'naturalness' of our relations to other people and things. How can one actively facilitate more episodes such as this in the workshops?

Negotiating didacticism

One of our main worries was that our workshops would be too didactic. In the very first *invisible Zürichs* workshop, with a school class of 16–17-year-old boys, we were surprised that there were almost no signs of imagination and play. The boys solved the tasks as if they were assignments given to them at school and were deeply concerned with doing it the 'correct' way. For example, when asked to find traces of the past, the boys meticulously documented old buildings and cars, factually trying to assess their correct age. The discussions centred around 'matters of fact', solely articulating a conceived representation of space, as defined by Lefebvre (1991). Space was here discussed in abstract terms, as a measurable geographic or empty geometric milieu, leaving little or no room for personal concerns. It was as if the 'proper' ways of engaging with urban space were replicated in the workshop itself. One boy had taken a picture of a man sleeping on a bench. When giving the picture a title, they were looking for the 'politically correct' term for the man: 'Would it be OK to call him a "penner" (slang for loiterer)?'

Certainly, we – the zURBS team – did not want the workshops to emphasize a certain 'correct' way of engagement. However, zURBS' curatorial choices were quite dominant in terms of deciding which places the participants would go to, what *Invisible Cities* stories they would read, and what tasks they would do. A constant discussion within the zURBS team was about the potential constraints of this control in terms of the creative freedom of the participants. However, whereas we liked the idea of no guidelines at all, we were aware that leaving the workshop completely open might also be experienced as constraining. For participants who did not normally engage in artistic projects such as these, the focus on 'everything goes' may seem intimidating, as the potential to reveal one's ignorance and/or lack of knowledge about events such as these is high (Bauman 2007). In order to minimize these potential constraints, we wanted to provide a clear framework so that the participants would not be insecure about what to do as they were sent out to explore the city on their own.

At the same time, we wanted to foreground a playful and imaginative approach that promoted pleasure and creativity, and this way enhanced the likelihood for shifts in identification. In order to encourage the participants to explore different identities and to act on other possibilities for being, we decided to set the scene by taking on roles as 'busy researchers of invisibilities and imaginations'. An inspiration in this regard was the work of performance artists such as Lois Weaver, whose alter ego Tammy WhyNot – a former famous country music singer turned lesbian performance artist – is used to avoid didactic lecturing and traditional narrative forms in favour of complex, playful, and welcoming theatrical experiences that address political issues from unexpected angles. At the beginning of each workshop, we staged a performance: I played the researcher-in-chief, commanding my minions to organize the archive, label its material, or restructure it (having a degree in acting, this was a part of the workshop that I enjoyed very much). As busy bees, Nina and Laura would buzz around, dusting, classifying, measuring, and inspecting the various items that filled the room. I would point out different objects for the participants of the workshop:

> Look at this ironing board! We believe it must be used as a bridge to cross a pedestrian path made out of floating water instead of asphalt. We are right now investigating how these pedestrian paths of invisible flowing water may provide new relations and patterns of movement in the *invisible Zürichs*. Or, take this box decorated with photographs. These are all images of things that mean other things. Like a dog that is a door and a fence that is a warning! We have a hypothesis that these are some of the significations of *invisible Zürichs*.

The performance attempted to encourage the imagination and facilitate a form of knowing that bears within it transformative force by letting the semiotic and experiential processes through which meaning is produced come into view. As the theatre director Anne Bogart points out:

> When we know what a door is and what it can do, we limit both ourselves and the possibilities of the door. When we are open to its size and texture and shape, a door can become anything and everything.
>
> (Bogart and Landau 2005, 59)

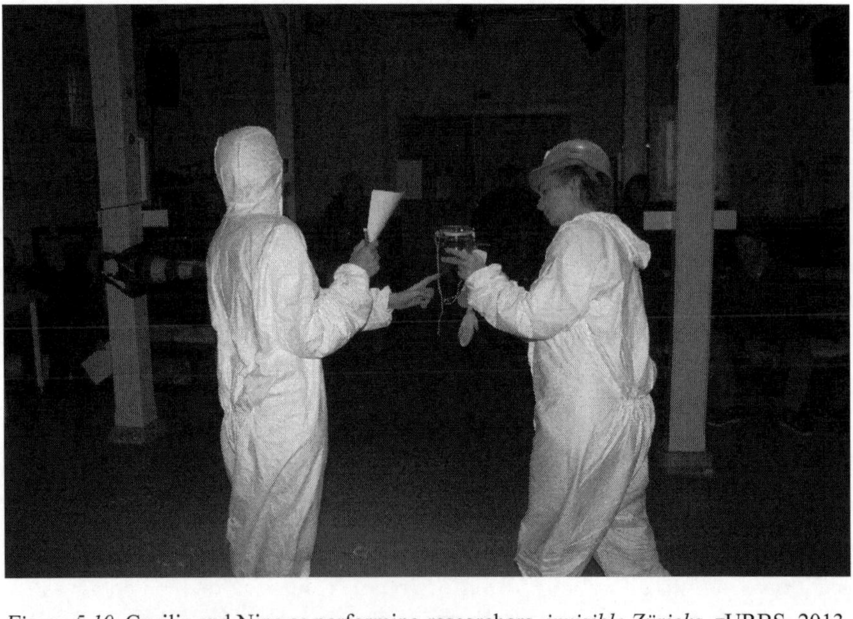

Figure 5.10 Cecilie and Nina as performing researchers, *invisible Zürichs*, zURBS, 2013 (image author's own).

Accordingly, the opening performance made explicit that the archive is given for its future (re)enactment (Derrida 1998). It does not preserve the past as past, but suggests a future for these pasts that has not yet taken place – much like a play-script in relation to production. This play-script is not a fixed, bounded set of conventions, but is produced in dynamic relation to attitudes and practices outside the playhouse. Hence, the subjectivities that emerged in the workshops were not autonomous, but highly dependent on social and material circumstances that were present outside of the workshop frame.

In describing his experience of the workshop, one of the participants, an architect in his 20s, pointed out that our costumes reminded him of 'Greenpeace or whatever activists'. This connotation was unintended from our side and made us change our costumes. Sabeth got the idea that the costumes should represent our different researcher roles. She would use her costume as a white canvas on which to write all her lists, as she is the one keeping track of all the practicalities of our projects. Nina filled her costume with tools and materials such as clips, a ruler, a magnifying glass, tape, plaster, cardboard, and so on, as she is the one conceptualizing and designing the 'scenography' of our projects. I filled my costume with quotes and printed pages from books and articles by theorists that I like, as I am the one always theorizing about our work. The development of these exaggerated versions of ourselves helped us adopt a playful attitude towards our roles as workshop facilitators. As Sabeth pointed out, she was much more comfortable with playing the role as an organizer than actually being it.

The participants' reaction to our initial costumes illustrates that representations (our researcher costumes) are inseparable from broader social practices that give them meaning (such as, for example, Greenpeace and activism). This presentational nature of the social and material world is more readily perceived in performance than in everyday life. Hence, performance may help expose the performative nature of all social and material relations, and accordingly enable actors to imagine acting differently from their normal sedimented patterns. A female student participating in one of the workshops pointed to how the workshops facilitated this process:

> For example, in the little garden, there were some moments when I felt intimidated, because the park was empty, there were only these two men walking and talking. And now that I think about it, I come from this country and city where you have to watch your bag all the time, you cannot trust strangers that easily, and I used to be like that [. . .] So at times I felt intimidated, but I saw the other two people with me, they were relaxed, they didn't care, so I was like: it is sad that I recognize it but I am in a different space now, so I do not have to watch my back as the place I come from.

In the beginning of *invisible Zürichs*, we wanted to give the participants specific tasks that would direct a certain form of transgressive behaviour and prompt them to act differently than normal. For example, if the participants were to take the tram to move between the neighbourhood where the workshop took place and the archive, we would give them the task to sit down next to a random person and ask where she or he was going. The idea was to have the participants have the bodily experience of breaking a social norm by sitting down next to a stranger and asking a personal question. As much as we liked the idea *in theory*, when time came to make it into a practical exercise, we decided against it. The reason was simple: the element of transgression should not be a task in itself, but rather a *means* for achieving something else. We did not want to create a situation where the participants felt pressed to try out some kind of social experiment and then come back to us and report their findings. This situation recalls the 'breaching experiments' conducted by Garfinkel and the psychologist Stanley Milgram. Both professors asked their students to carry out experiments involving them violating commonly accepted social norms and then examining people's reactions to it. Milgram, for example, asked his students to board a crowded train and ask someone for a seat. This seemingly simple task proved to be quite difficult, even almost traumatic, to carry out for his students. They reported feelings of nausea and fear, and most of them did not go through with the experiment. In line with our aim of facilitating positive affective registers, we wanted to avoid a situation where the workshops would generate negative experiences for the participants. Furthermore, we learned that shifts in identities cannot be 'known in advance or contained within a programmatic vision of how the world should be' (Cameron and Gibson 2005, 320). We could not and should not try to direct the behaviours of our participants, but rather provide a performative framework that would enable moments of transgression, new subjectivities, and unexpected encounters to emerge on the participants' own terms. This way, participants might or might not have the experience of crossing some line without necessarily intending it, and only then realizing that a boundary even existed.

Conclusions

In this chapter, I have discussed the different possibilities and challenges for residents of Zurich to 'take hold' of and engage in their urban environment. Due to official governmental control and regulation, as exemplified by poster campaigns such as 'Erlaubt ist, was nicht stört', urban space in Zurich could be seen as a 'natural order' with inherent meanings and predetermined functions that exercise control over the people who use it. The tension between proper use and personal use is here foregrounded. Through procedures of standardization, the so-called 'proper use' of space may serve as a justification for suppressing alternative uses of space as a 'disturbance' of public order and the common good. Accordingly, the city is alienated from the social life of its inhabitants.

invisible Zürichs attempted to question and counter this sense of alienation by focusing upon engaging the participants of the workshop in the articulation of personal concerns and attachments in relation to their city. This was done in two main steps. The first step was to send the participants into the streets in search for envelopes containing various tasks. The aim here was to facilitate a form of explorative engagement that would encourage the participants to change their perception of the city, break their normal routines, and accordingly act on other possibilities for being. The performative framework of this exploration sought to enable a form of transgressive behaviour that would make visible certain 'natural orders' of urban space by breaking them. This was, however, easier said than done. It proved difficult to engage participants outside of the normal 'culture crowd' in taking part in the workshops, as the abstract aim of 'changing the perception of the city' does not immediately relate to people's everyday concerns and occupations. Hence, even though we made several efforts at engaging a broad variety of participants, we had to admit that we could not include all, and that the workshops were subject to a certain 'exclusiveness'. Furthermore, we – the zURBS team – struggled with trying to avoid being too didactic in our approach. We realized that we could not control any of the outcomes of the workshop, for example in terms of whether any transgressive moments would take place.

In the next chapter, I discuss the second step in the process of engaging participants in their urban environment. While the first step involved engaging the participants in a playful and transgressive exploration of their urban environment, the second step relates to enabling the participants to articulate and communicate the personal experiences, memories, thoughts, and ideas that this form of explorative engagement evoked.

Bibliography

Bauman, Z. 2007. *Liquid Times: Living in an Age of Uncertainty*. Cambridge: Polity Press.

Bogart, A. and Landau, T. 2005. *The Viewpoints Book: A Practical Guide to Viewpoints and Composition*. New York: Theatre Communications Group.

Cameron, J. and Gibson, K. 2005. "Participatory Action Research in a Poststructuralist Vein." *Geoforum* 36 (3): 315–331.

Cresswell, T. 1996. *In Place. Out of Place*. Minneapolis, MN: University of Minnesota Press.

Derrida, J. 1998. *Archive Fever: A Freudian Impression*. Chicago, IL: University of Chicago Press.

Deutsche, R. 1988. "Uneven Development: Public Art in New York City." *October* 47: 3–52.

———. 1996. *Evictions: Art and Spatial Politics*. Cambridge, MA: MIT Press.

Edensor, T. 2001. "Performing Tourism, Staging Tourism." *Tourist Studies* 1 (1): 59–81.

Heidegger, M. 1962. *Being and Time*. Oxford: Blackwell.

Highmore, B. 2005. *Cityscapes: Cultural Readings in the Material and Symbolic City*. London: Palgrave Macmillan.

Lefebvre, H. 1991. *The Production of Space*, translated by D. Nicholson-Smith. Malden, MA: Blackwell.

Simonsen, K. 2012. "In Quest of a New Humanism: Embodiment, Experience and Phenomenology as Critical Geography." *Progress in Human Geography* 37 (1): 10–26.

Thévenot, L. 2014. "Voicing Concern and Difference: From Public Spaces to Common-Places." *European Journal of Cultural and Political Sociology* 1 (1): 7–34.

Urry, J. 2002. *The Tourist Gaze*, 2nd edition. London: Sage.

6 (Re)inventing urban democratic practices

Voicing matters of concern

After one hour out in the city collecting objects and other 'invisibilities', your group returns to the theatre stage with your findings. You are asked to put the material on a big table that everyone gathers around. Photographs are printed and placed alongside empty bottles of liquor, thistles and wet leaves, street magazines, dog poop bags, lost gloves, umbrella covers, a stack of small museum books . . .

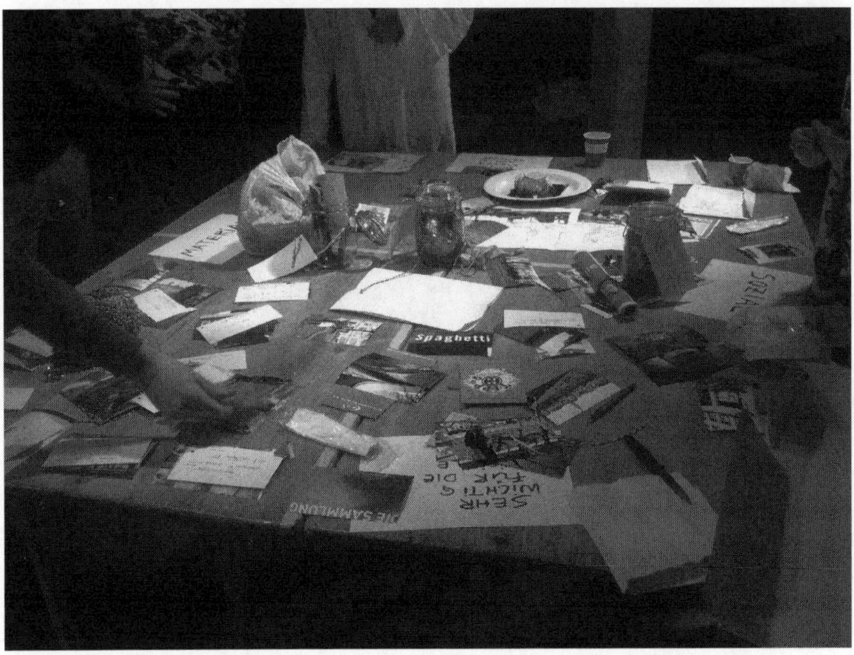

Figure 6.1 Objects collected by participants, 'the alternative city archive', *invisible Zürichs*, zURBS, 2013 (image author's own).

> People were really staring at us when we were heading back, because we were carrying all this weird stuff. In a sense that was nice also because we got this nice group feeling: in the frame of the workshop we had made this product without much introduction on what actually to do, and we had made it out of these easy things we found, like garbage stuff . . .
>
> (Hannah, 30s, HR worker, personal interview)

As requested, you label the objects that your group has brought back. You write down a title for the object, where it was found, and its story.

> I think [labelling is] one of the most important steps for the archive [. . .] It is something real, like a real finding. Like, before it was maybe a tomato or a bag full of plastic bottles, but with the label you make it official, like an official item of the archive. You give it a meaning, or your interpretation.
>
> (Roger, 40s, urban planner, personal interview)

Time has come to share the stories behind the findings. One group has brought a beverage coaster from a local pub and tells the story of how

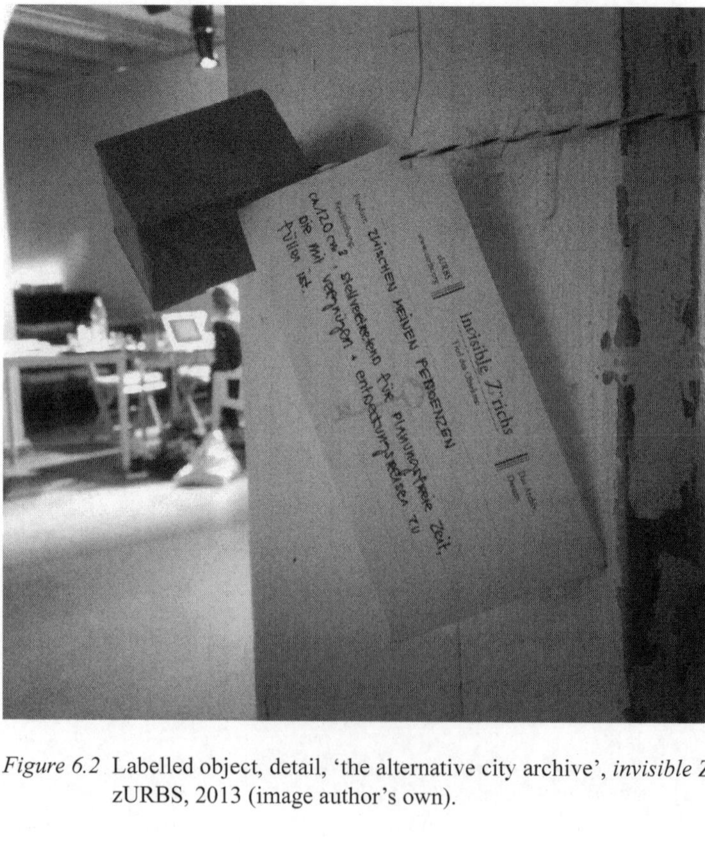

Figure 6.2 Labelled object, detail, 'the alternative city archive', *invisible Zürichs*, zURBS, 2013 (image author's own).

they discovered that both a woman and man in the group have, independently of each other, gone for a pint at this very pub after getting married in the city hall. A young student shows a photograph of a doorbell that says 'C. Inema' and tells that the collective that lives here regularly organizes public film-screenings for free. Another group has brought a box with luxurious cream-cake. You are offered to have a taste and you learn about this bakery that sells cakes from yesterday at a very reasonable price –sometimes cakes are even given away for free!

You are amused that two groups have brought back the exact same combination of two presumingly independent objects: a street magazine called 'SURPRISE' and a bag of thistles. The story is that both groups met this homeless guy that lives out of selling magazines in the street. If you buy a magazine, he gives you a bag of thistles that you can use like spinach.

Concerns about specific and peculiar places on the brink of extinction are expressed. One woman shows a picture of a display window from a shop selling navy uniforms, explaining that the house was for a long time the home of many squatters, but that it was now in danger of being torn down. As a reply, another woman tells about a nice little house that hosts a knitting club for retired women.

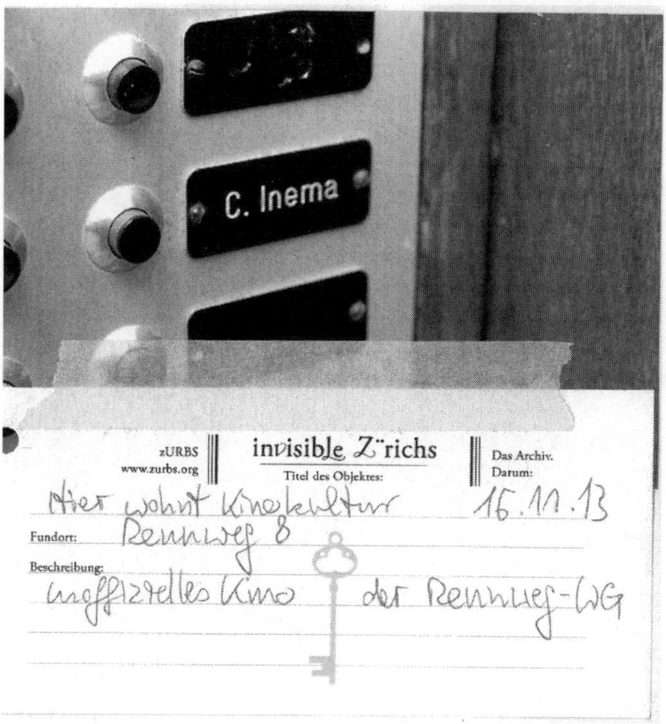

Figure 6.3 C. Inema doorbell, photographed by participants, detail, 'the alternative city archive', *invisible Zürichs*, zURBS, 2013 (image author's own).

This place has lost its support from the city and is also about to be torn down, she explains.

> I think the objects are kind of reflections of your souvenirs, but it is not your souvenirs themselves, but only a way, a security in order to write a narration and tell a story. So that you do not only write your story after the lived experience, but you rewrite the experience, and when you see it in here, in the archive, it is something new – so you write and rewrite something in the archive and the souvenir you have in the head.
>
> (Sara, 20s, student, personal interview)

The souvenir as generator of intimate narratives

Instead of relying on 'matters of fact', the workshops aimed to open up a form of communication based on 'matters of concern' (Latour 2005). 'Matters of concern' integrate differing expressions for voicing personal concerns, which in turn may create the conditions under which various residents can express, articulate, and create situated and shared knowledges and understandings of their city. Accordingly, we – the zURBS team – tried to facilitate an explorative process guided by an open form of storytelling that would enable the participants to articulate and communicate their stories of being in the city on their own terms.

This concern led us to focus on found objects, photographs, drawings, sound clips, scribbled stories, and samples of smells in laboratory glasses as

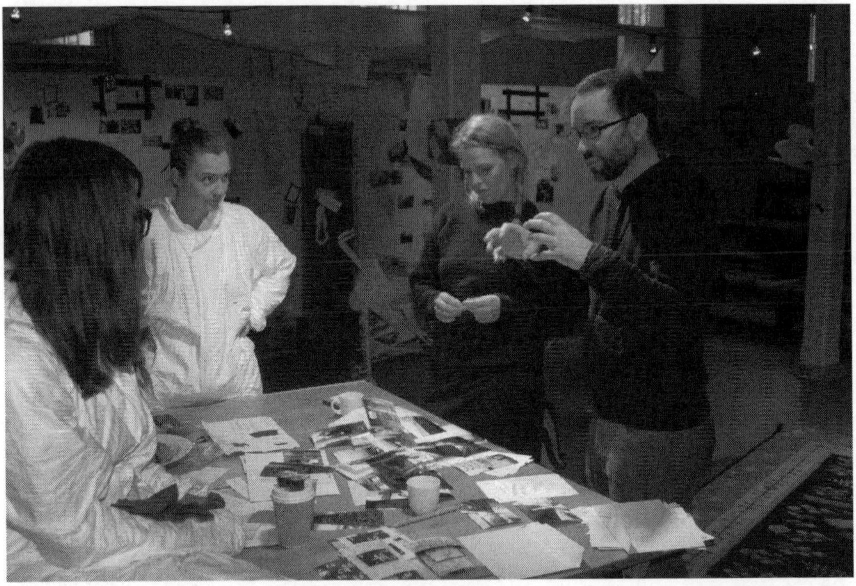

Figure 6.4 Participants sharing stories, *invisible Zürichs*, zURBS, 2013 (image author's own).

'communicative tools'. These 'tools' should help the participants to communicate personal concerns as well as bodily experiences and sensations that rely on a close and idiosyncratic hold on the environment. Thévenot (2007) points to the importance of people being able to 'communicate', not as a way of transmitting information, but as a way of participating in common matter (p. 412). Whereas transmitting information relies on a standardization that ensures that the information can serve as a source for anyone who might want to use it, communication reflects the many interrelated modes of engagement with, and experience of, our world. Thévenot calls for a move from information to communication in order to find ways of maintaining a composite and conflicting community that caters for the articulation of the variety of senses of reality in our lives and our living in common.

The expression of individual and personal opinions in Zurich is coloured by the idea of transmitting standardized information. The city's political decision-making process rests on a direct democracy in which the people are given the rights of initiative and referendum at all institutional levels. As John Gastil and Robert Richards (2013) point out, public issues are promoted alongside a single interpretation, denying the existence of alternative interpretations. People with differing views and interests here tend to add up to an 'unholy alliance' ('unheilige Allianz') (Hitz et al. 1993) that relies on either sacrificing personal views and concerns for a characterization of the common good, or formatting personal concerns into publicly available options designated as 'preference' or 'stake' or 'interest' (Thévenot 2014). The integration of differences is here achieved by negotiation between 'stakeholders', rather than letting various personal expressions find a common ground to communicate deep concerns, attachments, and feelings. Sabeth, Nina, and I wanted our workshops to be hospitable to these personal attachments and opinions by opening up a concept of urban space that enabled practical engagements below the level of standardized publicity and articulated personal engagements at a level of close intimacy.

The objects that the participants brought back to the archive can, in this regard, be seen as forms of souvenir. The souvenir, according to the poet and scholar Susan Stewart (1993), is an object that transforms and collapses distance into proximity, and thus contracts the world in order to expand the personal and enable a narrative of the self. The souvenir has the capacity to serve as a trace of authentic experience, such as the post-marriage pint mentioned in the workshop scene above. This is an event that is not repeatable, but it is reportable through the invention of a narrative. The narrative generates the souvenir, in this case the coaster, which in turn becomes the 'security' of this narrative. The souvenir, then, authenticates the experience of its owner/possessor and acquires its value through its material relation to the location of the experience.

However, the souvenir is always incomplete. It is removed from its 'natural' location, and thus is a mere substitution, sample, or trace of a now-distant experience that the object can only evoke or resonate with, but never entirely recoup. The objects brought back to the archive didn't make much sense to us until the participants told us the stories connected to them. It is exactly this partiality of the object (it will not function without a supplementary narrative) that is the source of its power. The souvenir is an allusion and not a model; it comes after the specific experience and location, and is thus only a part of, while at the same time also being more than, its origin. The street magazine and the thistles do not solely attest to the encounter with the homeless man; their supplementary narrative also creates a story of the hospitality and culinary skill and knowledge of this man. The encounter is thus expanded to a myth regarding the friendliness of the homeless people in Zurich, which in turn may counter the negative image of homeless people 'disturbing' the perceived natural order of the city. This narrative does not belong to the objects themselves, but to their possessors – the participants who met and talked with the homeless person. Hence, the souvenir is personal; it cannot be generalized to encompass the experience of anyone, only that of its possessor.

The souvenir, then, engages with the world on a level of close intimacy. As Stewart (1993) points out, the souvenir reduces the public to the miniature, to that which can be enveloped by the body, or into a two-dimensional representation, that which can be appropriated within the private view of the individual subject. A common reaction from participants entering the archive was of amusement at all the little details in it:

> It is very nice, I like it. It is like all these little details that you do not normally notice. You realize that the city is built up of all these little things. It is very nice with all these small stories. There is so many small details here that you would use hours to go through it. It is cool, it is a different way of discovering the city, like an un-normal way.
>
> (Lisa, 30s, student, workshop conversation)

> It is like there is not one overall community story about Zurich here, rather it is all these small stories.
>
> (Radena, 40s, student, workshop conversation)

The miniaturized world offers itself to possession and points to a transformation from the exterior to the interior, from public to private. As one visitor pointed out as she inspected the archive, 'It looks a bit like . . . like a living room. Just like in a living room you have all these pictures, stories and memories connecting to your own experiences'.

The analogy of the living room is interesting as it evokes the idea of an environment based on personal convenience. As Thévenot (2001) points out, the intimate familiarity of one's home is highly dependent on personal adjustments of one's familiar things, much like Wentworth's description of 'intelligence involved in practice'. For example, a missing door handle may be replaced by an adjustable wrench, a stack of books may support an uneven tabletop, or one may have developed a peculiar way of turning on the old kitchen stove. These forms of familiarization evoke a direct corporal implication in terms of a tight union between bodily gestures and the environment. Parallels can here be drawn to Heidegger's notion of 'dwelling', defining a way of inhabiting the world in a lived, experienced manner instead of one of abstract and calculative planning. Dwelling, then, is the direct opposite of a standardization that takes the city as a substance that can be ordered, planned, and worked upon – instead of worked with. The focus on the souvenir emphasizes that the city is not a spatial framework external to its users, but is produced by them. The idea that the city consists of a collection of neutral objects that speak for themselves is here countered; the city, like the souvenir, speaks through the narratives of its residents and users.

Producing counter-narratives

Your eyes scan the photographs that are lined up on the table:

- *a Styrofoam box that used to have oysters in it but is now filled with ice;*
- *a price list from a fancy clothes shop;*
- *the display window of an Asian street kitchen;*
- *the interior of a run-down telephone booth;*
- *a plastic mannequin tucked behind a tree;*
- *a sign with a picture of a watchdog;*
- *a street sign with the name 'Chaos platz';*
- *a sticker on a lamp post saying 'Hello my name is ...';*
- *two bikes locked together around a tree;*
- *a wall completely plastered with old photographs;*
- *colourful stones covering the ground of a neighbourhood park;*
- *a blindfolded man in a suit sleeping on a bench;*
- *a balcony with a colourful windmill on it;*
- *two empty seats with two hats opposite each other in the tram;*
- *a fence full of moss;*
- *a huge red ribbon hanging out of a window;*
- *an old gas pump;*
- *a close-up of a red typewriter;*
- *various fire hydrants looking like people;*
- *a shopping cart that has got lost in a garden;*
- *signs that say 'verbot' (forbidden);*
- *random people under a motorway bridge;*
- *a flowerpot with yellow flowers left on a bench;*
- *reflections of the sky in a car body;*
- *a door covered in posters;*
- *a pink bar chair leaning up against a garage door;*
- *the fence in front of a construction site;*
- *a window display full of hair extensions;*
- *a triangular shape that looks like a 'play button' lying on the sidewalk;*
- *a scarf hanging from a traffic sign;*
- *a buggy and a motorbike parked next to each other;*
- *platform no. 5 at Alstätten train station;*

- *parts of a facade with a blackbird painted on it;*
- *colourful chalk drawings on the pavement; and*
- *the lid of a green garbage can on which someone has scribbled 'Not to eat for humans!'*

You are amused that the idea of a clean and orderly Zurich is nowhere present in these pictures. Most motifs seem to focus on abject matter such as, for example, the greasy Asian street food, the dirty telephone booth, the hair extensions, and the garbage lid. These items play a central part in endowing the city with meaning by commentary, interpretation, and dramatic structure. The display of the Asian food is said to be an evidence of everyday artistry, the telephone booth is ridiculed for trying to provide a surveillance mechanism for the red-light district in which it is placed, the picture of the hair extensions has got the comic title 'Haare ziehen um' ('hair moving [house]'), and the garbage lid is seen as an example of how anything that is not perfect is not allowed to be used by humans again.

New images of Zurich are constantly formed in your mind as you listen to the stories told by the other participants. The picture of the empty tram seats is accompanied by a story about how people in Zurich almost become invisible

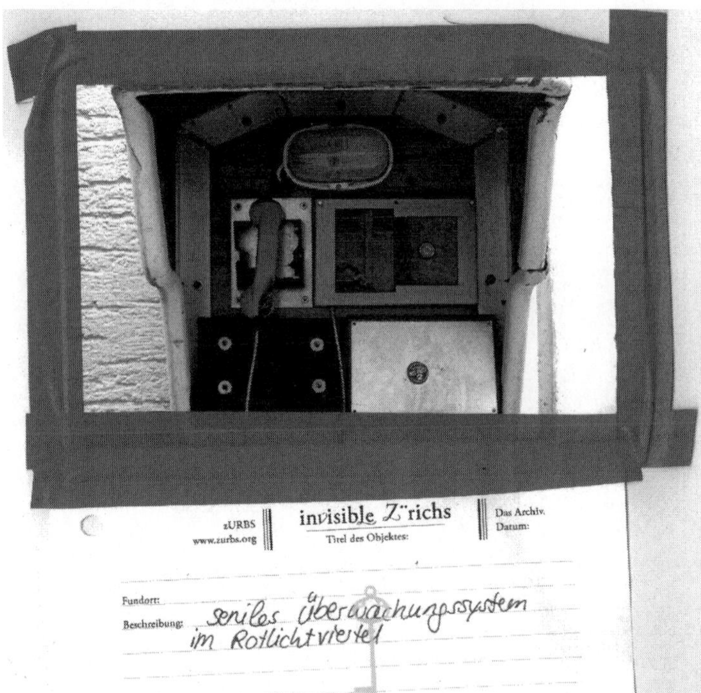

Figure 6.5 Telephone booth, photographed by participants, detail, 'the alternative city archive', *invisible Zürichs*, zURBS, 2013 (image author's own).

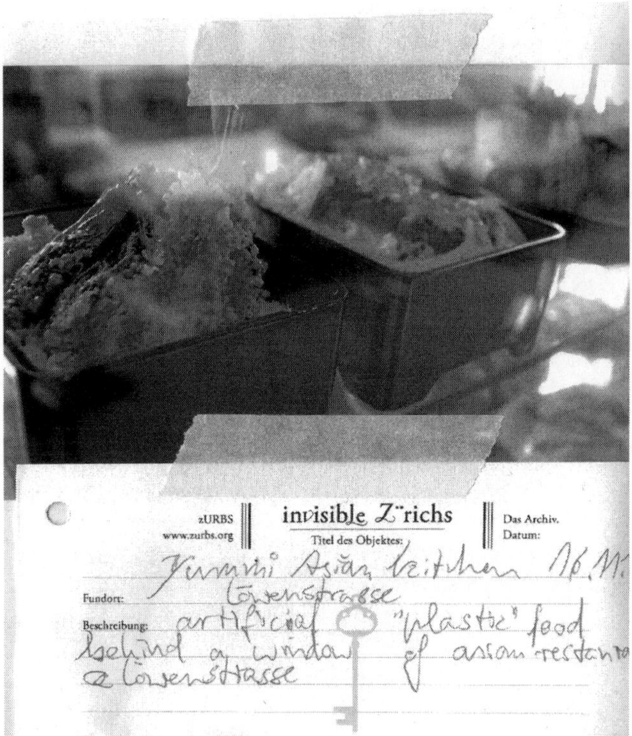

Figure 6.6 Asian street food, photographed by participants, detail, 'the alternative city archive', *invisible Zürichs*, zURBS, 2013 (image author's own).

in the winter. The female schoolteacher who took the picture explains that we wear so many clothes that we almost disappear. The clothes become an armour that protects us from our surroundings as we go about our own business and barely notice or communicate with each other. 'I wish it wouldn't be that way!' she sighs. The door covered in posters is accompanied by a childlike drawing expressing the desire for the door to lead to a fairy tale land where everything is free. The group explains how much they had enjoyed making the drawing because it had allowed them to dream of something else for a moment. The group that discovered the flowerpot on the bench tells that they have chosen to believe that it was left there intentionally as a little 'present' to the people attempting to sit on the bench. Therefore, the group decided not to bring the flowerpot itself to the archive, and rather took a picture of it, so that it could keep spreading 'the beauty' out in the city.

Maya: I am sure, when people see this archive, they would say that it is not Zurich. Zurich is not crowded and messy, they would say. But they are wrong; Zurich is really crowded.

Nidi: And I think this archive – made in this way – has to look like this. Like, the official archive in the city needs to be 'tidy', it needs to be categorized so that you can easily find what you are looking for. But this archive is not made like that, and so it has to look like this. I think any archive made like this, by the people of the city, has to be messy and chaotic.

A dispute about the colour of Zurich arises. A young woman has brought pictures of various facades in Zurich, arguing that Zurich has got 'such dull colours'. Some elderly ladies object, pointing to the beautiful yellow colour of one of the facades, to which the woman replies.

> But this colour is curry, it is not yellow, you know. It is still brownish . . . the red is also like . . . it is all greyish, it has this grey filter.
>
> (Helle, 30s, dentist, participant observation)

> And then there was me and [Joseph]. That was so funny because we have worked together for more than three years – he is the economist and I work in communication, with human stuff, but normally we understand each other sooo good. But at the workshop we did not understand each other at all. I had no idea we would have such different views on things [. . .] We had to understand each other in a new context that we were not used to [. . .] That was really interesting.
>
> (Hannah, 30s, HR worker, personal interview)

The facilitators announce that it is time to place the material in the archive. You are free to arrange it however you want; hang it from the ceiling, plaster it on the walls, place it on the floor. However, they encourage you to place it in relation to the things that are already there, in order to comment upon it, challenge it, or complement it.

A picture of a park with a statement warning that it is not suitable for families, as the ground contains shiny colourful stones that children are prone to eat, is counterpoised with a statement referring to the beauty of the park, advocating for the importance of having parks such as these in Zurich; a story based on a photograph of surveillance cameras, referring to the feeling of being constantly controlled, is juxtaposed with a story of the loneliness of the city, due to the predominantly blue colours in the photograph.

The workshop is coming to an end. The facilitators thank you for your time and effort and encourage you to keep looking for invisibilities in Zurich. You are told that you can come and visit the archive any time in order to follow its development as more workshops are executed.

> I was proud having my personal contribution as part of [the archive]. Now I could look at all of these things and go: my contribution is also there somewhere in all this. What I liked the most with the workshop is that it depends so much on yourself what you get out of it. You really have to invest of yourself.
>
> (Hannah, 30s, HR worker, personal interview)

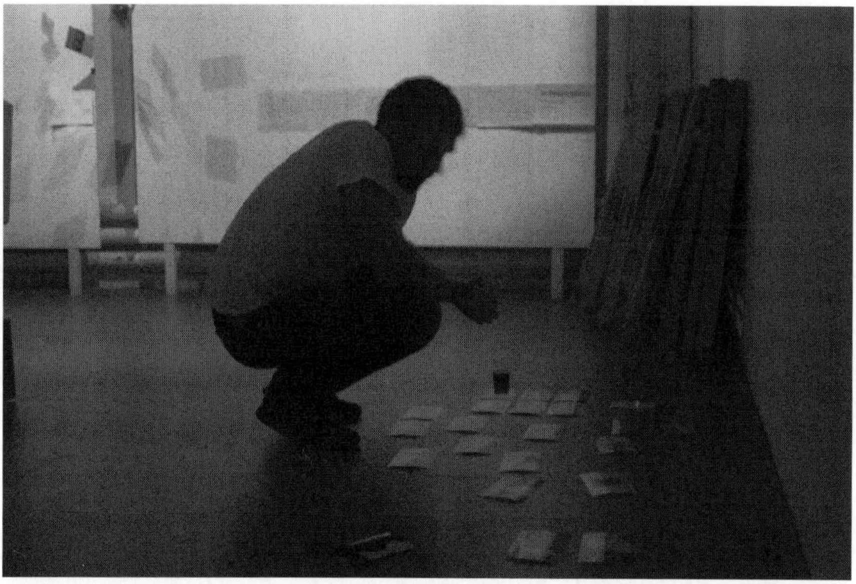

Figure 6.7 Archive-making, *invisible Zürichs*, zURBS, 2013 (image author's own).

Reframing the abject and the beautiful

The participants of the workshops had not taken pictures depicting panoramas of important and extraordinary buildings and places in Zurich, such as the renowned and glamorous shopping street, Bahnhofstrasse, the picturesque Lake Zurich, or the landmark site of Grossmünster's two church towers. Instead, they had taken pictures of what Perec (1997) defines as the 'infraordinary' – that which lies beneath notice or comment, 'that which we generally don't notice, which doesn't call attention to itself, which is of no importance' (p. 46). A significant number of these pictures depicted 'messy' details and traces of use that focused on 'disorderly' and 'unclean' aspects of urban space.

This is interesting in terms of how traces of use are perpetually and immediately erased in Zurich. Wentworth uses the word 'sanitation' to describe the approach to urban space in the city: 'I think "sanitation" is the right word, it is not "hygiene" or "cleanliness"' (personal interview, 21.10.2014). This choice of words is particularly interesting coming from an artist who builds his work around a direct contact and engagement with urban space. 'Sanitation' points to the prevention of human contact with the hazards of wastes, and thus evokes a feeling of distance rather than of proximity, connection, and sensation relating to touch. As Thévenot (2001) argues, a familiar and close engagement with one's surroundings has a direct corporal implication, the idea of a tight union between bodily gestures and an environment that makes for highly local convenience. 'Sanitation', on the contrary, can be seen in relation to what Davina Cooper (2014, 71) observes as a governing through vision. This form of governing appears panoptical, efficient, clean, asymmetrical, and fast. For example, every year Zurich hosts one of the largest technoparades in Europe, the Street Parade. Although more than half a million people throw a giant party in the streets, the next morning any remnants of confetti, plastic glasses, and bottles are meticulously removed. This immediate removal of traces of use points to a negative view that is not solely oriented around removing signs of use due to their negative connotations of deterioration or decline. It also relates to the uneasiness caused by the absence of any indication that someone is working hard to remove traces of use, whether it is the smell of urine on a street corner or the remnants of a half-eaten kebab. The mistrust caused by these signs of wear is often related to how other users are here forced to associate themselves with wear that disturbs them (Boniver et al. 2010).

Julia Kristeva's theory of 'the abject', referring to 'what disturbs identity, system, order' (cited in Campkin 2013, 56), is worth mentioning here. Abject matter can be seen as a combination of physical, moral, and psychological reactions, such as, for example, feelings of disgust at a rotting object or of moral repulsion to a horrific crime (Cox and Campkin 2007). Kristeva's theory builds directly on Mary Douglas' (1996) seminal anthropological text *Purity and Danger*, in which the latter conceptualizes dirt as 'matter out of place'. Just as the environment can be depicted as having a 'natural' order, Douglas points to how dirt appears to be defined through an intuitive and natural belief based on bodily sensations and psychological processes. However, as Ben Campkin (2013) points out, there is no such thing as absolute dirt – it is a matter of perception and classification. Hence, the city can be seen as a place where we are continuously forced to negotiate the dirty and abject. Within this context, abjection can be seen as spatialized processes through which the subject, or society, attempts to impose or maintain a state of purity.

The 'messy', 'unclean', and 'disorderly' details that were depicted in the participants' photographs can be seen as a negotiation of this abject matter. The photographic capturing of these motifs gives them a certain solid and justifiable presence that produces a counter-narrative to the story of the conventional, clean, and orderly Zurich. The abject is here not seen as 'matter out of place'. On the contrary, it was frequently linked to what the participants defined as positive and beautiful aspects of urban space.

This 'aestheticization of the abject' in the workshops should not be confused with Andrew Harris' (2012) notion of 'the urban pastoral'. 'The urban pastoral' refers to the portrayal of simple yet virtuous representations of everyday artefacts and abjection of working-class life in urban settings. This portrayal relates to the traditional idea of the pastoral as rooted in poetic depictions and representations of an idealized, 'simple' rural life created for and by a socially and economically more advantaged audience with a particular attitude and perspective. Accordingly, Harris points to how 'the urban pastoral' can be seen to produce and reimagine urban space for more affluent social groups. Post-industrial degradation and inner-city environments featuring dirt, dereliction, and graffiti are here sparking off a pastoral fantasy in which artists celebrate objects, practices, and histories of marginalized social groups, while keeping a 'safe distance' by not directly engaging with the material conditions of these groups.

The participants in *invisible Zürichs* did not romanticize the abject in a disinterested 'urban pastoral' manner. They regarded the abject as beautiful because it was directly linked to personal everyday interests. Hence, the participants refrained from framing the beauty of the abject as belonging to some form of aesthetic uniqueness free from the interests of politics, status, and daily life. Rather, the abject was considered beautiful on the very basis of its relation to everyday social practices. The workshops, then, countered aesthetic strategies in which beauty is claimed to be something universal that lies outside of socio-material conditions.

As Thompson (2009) argues, beauty is not self-contained, but depends on a wider context of judgements. We feel compelled to share a response to what is perceived as beautiful, demanding assent from others regarding it in order to ensure that the individual experience becomes communal. Beauty, then, can be seen as 'a stimulus to collaborative work – it is an invitation to participate' (Thompson 2009, 144). We see a beautiful object and the feeling is so precious that we are encouraged to reproduce it; to hold, photograph, draw, or describe what we have seen so as to 'validate' our experience and also share with others the same sense of pleasure. This quality of beauty demands a personal engagement based on subjective perceptions, positions, and understandings. The idea of universal claims is here negotiated and challenged, as the constructedness and diversity of beauty are put to the fore. This process may generate a new openness towards the material world and draw people into it. Hence, the urge for discovering, creating, and sharing beauty may increase our sense of interconnectedness, being in the world, and sense of presence. For example, one participant, a student, told about how an elderly woman in her group got very touched when reading about the city of Phyllis as they were in a backyard surrounded by doors in all kinds of colours. The description of Phyllis emphasizes the joy of observing all the bridges over the canals and their differences, the variety of windows that look down on the streets, the many kinds of pavements that cover the ground, and so on. The elderly woman had immediately spotted the majority of blue doors that surrounded the backyard and made a drawing depicting only these blue doors and nothing else – as a response to the text. The student explained how this for her was the most beautiful moment in the workshop as the drawing was not only making the beauty of this specific place evident to her, but because the drawing was resonating so much with the picture of Phyllis that she had in her own mind after reading the story.

This example points to how the elderly woman and the student entered into relation to each other by sharing something that they experienced as beautiful. It was particularly important for the student that her idea of beauty was confirmed by the elderly woman's drawing. The interrelationships and interdependence with others are here favoured over our own autonomy. Accordingly, we are motivated to start a debate and negotiate what is beautiful, what is good, what do we like, what don't we like. This is illustrated in the little dispute about the colour of Zurich, as described in the workshop scene above. The negotiation of beauty is here related to the negotiation of the abject, showing that what is beauty for some is abject for others. As these notions are not possible to define, we know that we cannot insist upon our own meanings and perceptions in our claim to make them universal.

This way, the workshops countered the role of 'beauty' as committed to an aestheticization of the city that actively aims to remove perceived 'disturbances' such as waste, noise, and homeless people through processes of standardization, regulation, and control, as well as the more passive form of 'the urban pastoral' that promotes a disembedded aesthetic disposition that simply does not engage with these realities of urban space. Rather, the participants evoked a notion of beauty that pointed to an urban environment receptive to a personal engagement which is inseparable from the 'messiness' that is either glossed over or deemed out of place by the city authorities. This reframing of the concept of beauty produced a political space, alive to subjective assertions as well as public negotiations as to what is beautiful and right, or abject and wrong. This space makes explicit the tension between personal specifics and universal claims, and asserts that any universal claims about what is beauty or what is abject cannot command obedience. However, the openness of this approach is vulnerable to critique. If universal claims have no foundation other than personal and individual interpretation, don't we then risk falling into the trap of relativism and denying responsibility and critical inquiry?

Troubling aesthetic evaluation

In a workshop with employees and users from an organization providing a meeting place for alcoholics, I noticed how participants would use the argument 'this is art' ('Das ist Kunst!') repeatedly when they were making something for the archive, and I asked them to explain what it was. The participants of this workshop seemed to have quite a different perception of what art is than I had. Whereas I considered the artistic perspective offered through our workshops to be liberating in that it would enable the participants to articulate the experiences of the city on their own terms, the participants perceived the liberatory potential of art to be exactly the opposite: that they did *not* have to articulate themselves. This perception of art, as what does not have to be explained, can be seen in relation to public discussions concerning art in which politicians, artists, and art commissioners often talk about the importance of, for example, public artworks, but refrain from explaining the artworks as such (e.g. see Zebracki 2012). The popular view that art cannot be defined is relevant here. Artists, art educators, and art audiences alike hold to this view, grounding it in the observation that the things we call art do not have a distinctive feature in common.

This vocabulary of art provided the participants with a certain democratic freedom in relation to the political space created in the workshops. For zURBS, this political space was based on a rethinking of what is given. We understood art as a tool to refuse to know your place, to turn things upside down and inside out, in order to provide new perspectives on the world and take politics out of the realm of the taken for granted (where everything is in its place). While praising the openness of this approach, we failed to recognize its constraints. Our view on the relation between art and politics was one very much based on a particular mode of aesthetic valuation. In line with the surrealist approach, we expected the artistic framework to surprise us with unlikely choices and new perspectives. Bourdieu (1984, cited in Binkley 2000) points to how this mode of aesthetic valuation belongs to a so-called bourgeois or aristocratic taste that competes with the working-class taste, which expects a practical, sensual correspondence between content and form: beautiful art should depict beautiful things. This form of 'taste hierarchy' (Binkley 2000, 150) may provide an insecurity in terms of how one's aesthetic valuation relates to the individual sense of self, and one's sense of place within a social framework. By creating this very particular aesthetic framework, did we – the zURBS team – in fact produce our own urban hegemony?

Kester (2004) is wary of artistic practice in which a strong artistic authorship leads to a situation in which the artist is seen as the privileged bearer of insights, patronizingly informing audiences as to 'how things really are'. Bishop (2012), however, claims that this argument 'seems to present the participants of collaborative art as dumb and fragile creatures, constantly at risk of being misunderstood or exploited' (p. 26). As I will discuss in the next chapter, we did not want to judge in advance what people were capable of coping with and censor ourselves and the participants. No doubt, zURBS' aesthetic approach and artistic authorship did limit the workshops in certain ways. If we, for example, had produced a community garden rather than an archive, it might have been easier to engage participants who were unfamiliar with our approach. However, transgressing one's habitual comfort zone was an important part of our workshops, in order to break with routine perceptions and habits. Our focus on storytelling offered the participants models for agency, transformation, and change in this regard. The workshop with the employees and users from the meeting place for alcoholics illustrated this. For me, this was the workshop that created one of the most beautiful and poetic contributions. The hubcap of a car wheel was used to create a mobile, from which the participants hung photographs of trash in the city. In making the mobile, the participants engaged in linking the conceptual with the material realm of urban space. Much thought and discussion were given to how the photographs would be best (re)presented. This pointed to the participants' concern with mediating their concrete experiences of urban space, as depicted in photographs, through a mental construction, or conceptualization, of these perspectives. In other words, the participants appropriated the symbolic production of urban space in Zurich. I am cautious to make claims about the potential for transformation and empowerment in this form of appropriation. However, I do believe that in actively engaging in this process, the participants were enabled to reflect upon their own roles as inhabitants and their potential to act as 'stewards of public space', to repeat van Heeswijk's term. Accordingly, the creation of the mobile provided the participants with a form of 'security' by being rooted in the conventional and practical metres of everyday life, and thus transgressing the insecurity of aesthetic evaluation.

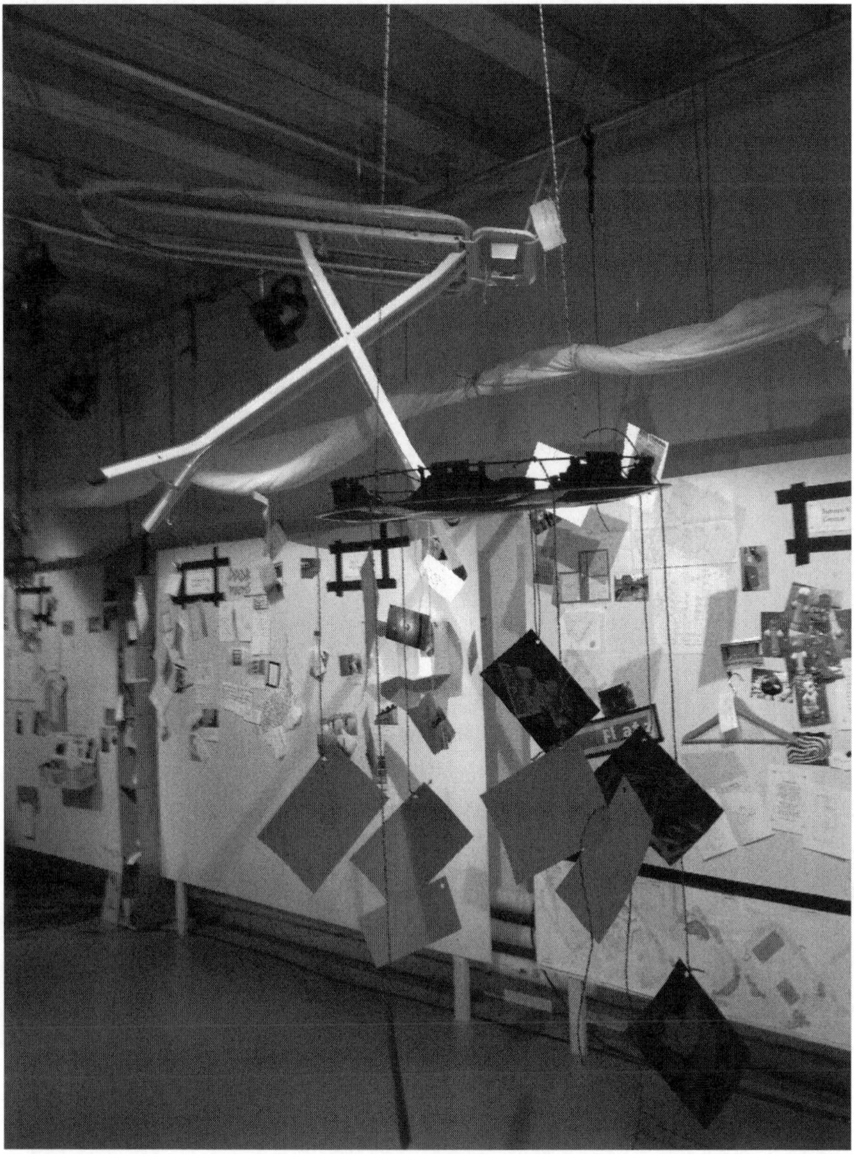

Figure 6.8 The hubcap of a tyre made into a mobile, detail, 'the alternative city archive', *invisible Zürichs*, zURBS, 2013 (image author's own).

Epilogue: the future of the archive

As *invisible Zürichs* was coming to an end, the question 'What will happen to the archive when this project is over?', frequently asked by the participants, became pressing. It was important for us – the zURBS team – to show the participants that we took seriously their contributions and the time and effort they had spent in finding, gathering, and developing material for the archive. The initial idea for 'ending' the project was to have an official ceremony in which we would hand over the archive to the city of Zurich. However, we did not succeed in getting any representatives from the city on board with the idea. Furthermore, the ceremony would only be an 'empty' ritual, as the city would never actually take over the archive, and so all the material would inevitably end up in the dump. Hence, we finally decided to invite artists that we knew and that had been following the workshops to turn the material into artworks based on their own interpretations of the archive. Important for these artworks was to 'take the archive further', to show that it was not just of the past, but a 'basis on which we *remake worlds*' (Yusoff 2008, 7, original emphasis) for the future. We gave the artists the task of creating artistic interpretations for how all this archived material could point to a possible future for Zurich.

Six 'artworks' came out of this task. Nina made the installation *Open It*, in which she scanned all the jars in which the participants had 'conserved' elements of Zurich. By scanning tactile objects, Nina questioned the digitalizing of archives and the priority of pictures and text-based documents in much archival practice. The resulting scans explored the interface between what is lost when turning objects into printed documents, and also what can be gained. Sabeth, in collaboration with Dominik Fornezzi, built a labyrinth out of the cardboard boxes, labels, written notes, pictures, and objects that constituted the material of the archive. The labyrinth ended up as a cave-like construction with a room for two people to sit down in the middle of it and have a look at all the material surrounding them. Here, a so-called 'what-if hat' was hanging from the ceiling. Visitors were invited to put on the hat and use the flashlight hanging beside it to light up the different things they would want to investigate closer. Brandon Farnsworth, an art student at Zurich University of the Arts, used what he described as 'randomizing processes' to draw two sets of five totally unrelated items from the walls of the archive. By 'curating' a mini-museum of these elements, Brandon wanted to explore the multiplicity of meanings hidden within the relationship between the archive and viewer, while attempting to overcome any trace of his own subjectivity. Another art student, Mirjam Wirz, selected written words on the different documents and material of the archive, as well as words she had heard spoken in the archive. She separated the words from their contexts and grouped them into what she defined as so-called 'empty' words (*verbrauchte Wörter*) and words with potential (*ausbaufähige Wörter*). It turned out that the words she read in the archive mostly belonged to what she considered words with potential, while the words spoken were mostly situated in the group of empty words. Nevertheless, these categories overlapped as Mirjam observed that the words appearing in the texts were frequently repeated in the conversations appearing in the room. The artist Ildi Serako made an installation of a warm winter coat hovering over the archive. The inside of the coat was filled with pictures from the archive. Ildi wanted to depict the kindness and warmth of the city and its memories, how it may protect us from the cold, and also how it may lift and guide us into an unknown future. Finally, I made a colour scheme out of the pictures in the archive. The scheme pointed to the different colours Zurich may take on, depending on personal perspectives, moods, and times. We invited all participants of the workshops, as well as other contributors to the project, to a final vernissage of the artworks and a following closing party of *invisible Zürichs*. The workshop participants seemed happy about

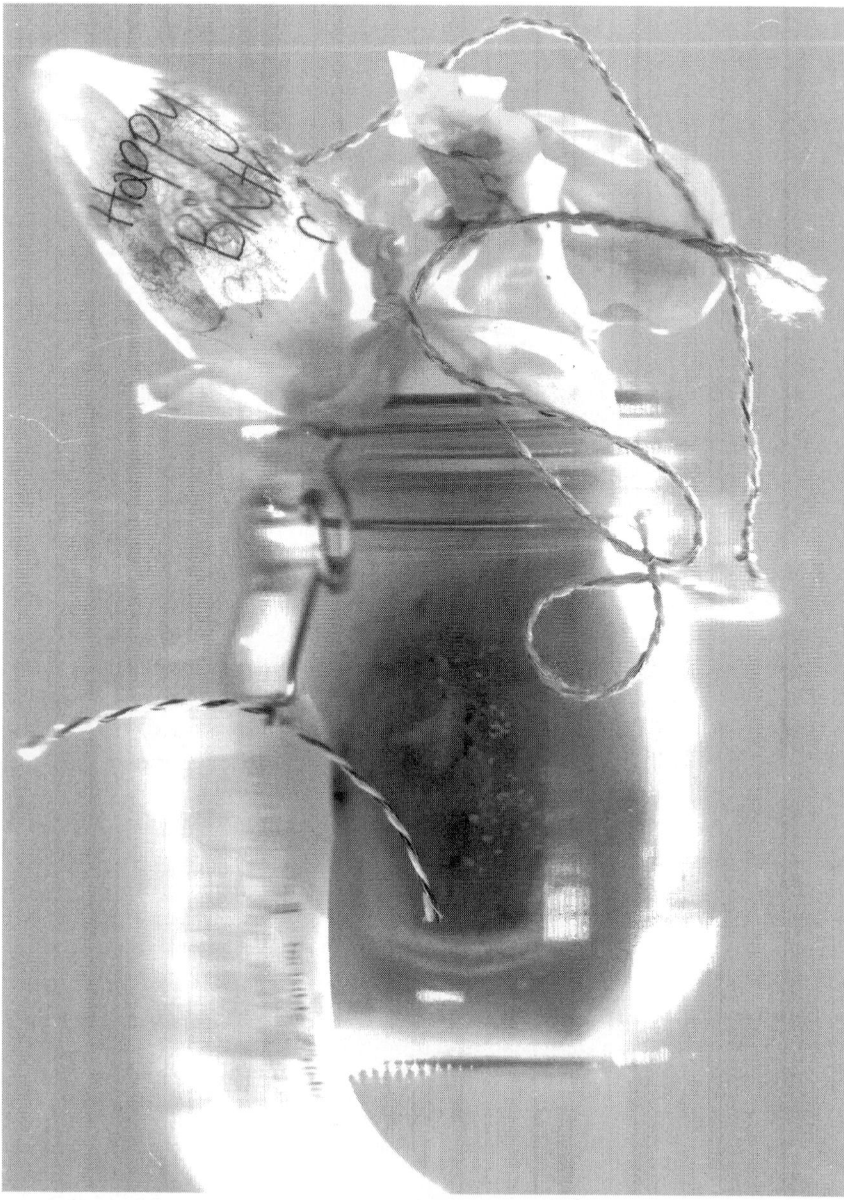

Figure 6.9 Detail, *Open It*, Nina Lund Westerdahl, 2013 (image author's own).

the transformation of the archive, searching the artworks in order to find their personal contributions that yet again had taken on a new meaning.

Conclusions

invisible Zürichs points to the complexities socially engaged art faces when trying to facilitate participative processes in which participants are encouraged to critically engage with their everyday urban environment in order to recognize, question, and challenge the material and social circumstances that regulate their everyday actions and behaviours. In order to articulate and communicate the personal experiences, memories, thoughts, and ideas that the explorative engagement with urban space evoked, the participants used found objects, photographs, drawings, and scribbled stories as 'communicative tools' and souvenirs that acted like a 'security' for the authenticity and relevance of the stories told. Even though it turned out that, for example, the 'souvenirs' got new meanings assigned to them when placed in the archive, personal intent and the authority of the items were still highly present. Hence, the archive emphasized that the city does not consist of neutral objects that speak for themselves, but that the city speaks through the narratives of its residents. These narratives countered the idea of a clean and orderly Zurich by focusing on the 'infraordinary', disorderly, unclean aspects of urban space. Rather than looking at this 'abject matter' as a disturbance, the participants emphasized the beauty of these elements, and thus negotiated definitions as to what constitutes the abject and 'matter out of place'. This way, the workshops opened up practices of urban democracy in which taken-for-granted assumptions of the 'proper' use and definition of urban space were negotiated. This political potential was further enhanced by the presentational aspect of the workshop, exposing how all materiality is (per)formed through the social practices that give them meaning. Accordingly, the workshops attempted to make the participants aware of how their actions are not solely influenced by urban space, but that their actions may also change urban space, and thus also how they act in it.

However, through these processes, we became aware of a series of challenges and contradictions within our practice. On one hand, we trusted our performative framework to provide an assumed 'free' space that would enable the participants to engage with the structure we provided on their own terms. On the other hand, our workshops were framed by a particular aesthetic approach that might very well have compromised the participants' sense of ownership by favouring a certain form of engagement. This contradiction raises the question of the forms of participation that are enabled through socially engaged art. I will examine this question in the next part of the book, which revolves around the zURBS projects that followed *invisible Zürichs*. Whereas this part has provided a practical perspective on the relation between socially engaged art and the politics of urban space in Zurich, the next part will have a closer look at participation in relation to the space created in and through socially engaged art itself. The representation of people and their interests is here in focus.

Bibliography

Binkley, S. 2000. "Kitsch as a Rpetitive System." *Journal of Material Culture* 5 (2): 131–152.

Bishop, C. 2012. *Artificial Hells*. London: Verso.

Boniver, T., Devlieger, L., Ghyoot, M., Gielen, M., Lasserre, B., Tamm, M., and Zitouni, B. 2010. *Usus/Usures: How Things Stand*. Brussels: Snel.

Bourdieu, P. 1984. *Distinction: A Social Critique of the Judgement of Taste*, 8th edition. New York: Routledge & Kegan Paul.

Campkin, B. 2013. "Placing 'Matter Out of Place': Purity and Danger as Evidence for Architecture and Urbanism." *Architectural Theory Review* 18 (1): 46–61.

Cooper, D. 2014. *Everyday Utopias: The Conceptual Life of Promising Spaces*. Durham, NC: Duke University Press.

Cox, R. and Campkin, B. 2007. *Dirt: New Geographies of Cleanliness and Contamination*. London: I. B. Tauris.

Douglas, M. 1996. *Purity and Danger*. London: Routledge.

Gastil, J. and Richards, R. 2013. "Making Direct Democracy Deliberative through Random Assemblies." *Politics & Society* 41 (2): 253–281.

Harris, A. 2012. "Art and Gentrification: Pursuing the Urban Pastoral in Hoxton, London." *Transactions of the Institute of British Geographers* 37 (2): 226–241.

Hitz, H., Schmid, C., and Wolff, R. 1993. "Urbanization in Zurich: Headquarter Economy and City-Belt." *Environment and Planning D: Society and Space* 12: 167–185.

Kester, G. 2004. *Conversation Pieces*. Berkeley, CA: University of California Press.

Latour, B. 2005. "From Realpolitik to Dingpolitik." *Making Things Public: Atmospheres of Democracy*, 14–44. Available at: www.bruno-latour.fr/node/208 [accessed 23.10.2018].

Perec, G. 1997. *Species of Spaces and Other Pieces*, 2nd edition. London: Penguin.

Stewart, S. 1993. *On Longing*. Durham, NC: Duke University Press.

Thévenot, L. 2001. "Pragmatic Regimes Governing the Engagement with the World." In *The Practice Turn in Contemporary Theory*, edited by T. R. Schatzki, K. K. Cetina, and E. von Savigny, 64–82. London: Routledge.

———. 2007. "The Plurality of Cognitive Formats and Engagements: Moving between the Familiar and the Public." *European Journal of Social Theory* 10 (3): 409–423.

———. 2014. "Voicing Concern and Difference: From Public Spaces to Common-Places." *European Journal of Cultural and Political Sociology* 1 (1): 7–34.

Thompson, J. 2009. *Performance Affects*. London: Palgrave Macmillan.

Yusoff, K. 2008. "Archive Fever." In *Bipolar*, edited by K. Yusoff, 6–7. London: Arts Catalyst.

Zebracki, M. 2012. "Engaging Geographies of Public Art: Indwellers, the 'Butt Plug Gnome' and Their Locale." *Social & Cultural Geography* 13 (7): 735–758.

Part IV

Representative frames

Constituting identities and issues in socially engaged art

Representation as a constitutive act

In Part I, I discussed the issue of the politics of representation regarding the socially constructed nature of meaning, and the need for a careful negotiation of how to mobilize and represent communities. Representation is here defined as someone standing for or acting for someone else, as well as the making of depictions or portrayals of others. To act for someone is unavoidably to portray her or him in a certain way. The conclusion of this part foregrounds that it is not sufficient for socially engaged art to represent various perspectives and multiple worlds. Additionally, it is necessary to recognize the opportunities and limits within our environments to enable or constrain the creation and representation of these various perspectives and worlds. The previous two chapters focused on these opportunities and limits in relation to *invisible Zürichs* and the politics of urban space in Zurich. I illustrated that the city does not speak for itself, but through its inhabitants. These inhabitants are not a homogeneous social group, and their ability to speak, as well as what they have to say, depends on a variegated politics of identity and difference. Whereas I here discussed how socially engaged art can provide a space in which a diverse set of identities and interests relating to the city could be created, expressed, and negotiated, this part will focus on how this space – as a framework for representation – may enable or constrain the production of these identities and interests. This focus relates to my third set of research questions, which centres on the particular frames of representation provided by socially engaged art, and the kinds of participation, subject-positions, and possibilities for self-representation that are enabled within these frames. Just as artists may seek to represent things by creating objects, so socially engaged art represents its participants by portraying them and their interests in particular ways. Accordingly, the political scientist Michael Saward (2005, 2008, 2010), in his theory of 'the representative claim', argues that the makers of claims constitute or create the object of their claim in the process of representing these claims. Representation is here understood in two senses: both to make claims and also to construct these claims.

Representation, then, is central to active citizenship – what Lefebvre terms 'autogestion', referring to 'self-management' or 'self-government' (Elden 2004).

Lefebvre argues that inhabitants have the right to participate centrally in the decisions that produce urban space. This right is not God-given, nor natural, but the end-result of collective claims made by mobilized citizens (Purcell 2014). In making these claims, citizens mobilize and initiate a struggle to do away with the distance between the institutional powers of the state and citizens in civil society, and the resulting homogenizing powers that forcibly classify people in predetermined categories. As Saward (2010) observes, most accounts of representation have an unfortunate tendency to presume that the represented have a given, transparent, and largely stable set of interests. The role of representation needs to be taken more seriously in terms of how it constitutes identities and issues, rather than merely reflects pre-existing ones. Hence, Saward argues for attending to the performative aspects of representation. Rather than treating it as a fact or thing, representation is here seen as an 'event' or constitutive activity that involves offering constructions or images of constituents to constituents and audiences.

The performative aspects of representation are highly relevant in discussions around the participatory aspects of socially engaged art. They point to the critique that this practice tends to obscure the discursive construction of its participants, as mentioned in Part I. Accordingly, this part will examine how a politics of representation is practised and reshaped within the context of socially engaged art. The ubiquitous aspects of representative politics are here foregrounded; people and things are not invested with meaning without representation. Issues of representation are thus central to further examination of how socially engaged art may challenge conceptions on the production of urban space within the neoliberal city.

Difficulties and inherent contradictions

The concept and practice of representation have recently received renewed critical attention. Concern about the marginalization of certain groups from representative structures and widespread disaffection with representative politics has resulted in a focus on a deliberative democracy (Dryzek 2000), focusing on reasoned argument and public reflection as much as voting and aggregation. Questions concerning who gets to deliberate, where, and how, here emphasize that representation is always tenuous and difficult to achieve convincingly (Saward 2008).

This difficulty was something we – the zURBS team – bitterly experienced when organizing and executing *invisible Zürichs*. We believed that we could provide a so-called 'untainted' (Saward 2005) space for representing marginalized interests because we were doing an 'art project' independent from any political institution or party. This belief relates to an understanding of the 'aesthetic' as a distinct, universal, and assumed 'free' space that we access through art (Bishop 2004, 2012a, 2012b). However, socially engaged art runs into the problem that this putatively 'free' space of art is constituted, or at least surrounded, by practices that reinscribe social divisions. For example, while attempting to provide a performative frame in which the participants could tell their own stories of the city, we also heavily authored the workshops in terms of giving the participants specific tasks, routes, and focuses. The participants were not looking for *any* aspects of Zurich, but distinct *invisible* aspects of the city. Were we, in fact, encouraging participation in order to make people think more like zURBS? And if so, what was the emancipatory potential in that?

We increasingly had the feeling that *invisible Zürichs* operated within an aesthetic discourse and political agenda that for some appeared both abstract and exclusive. As Sabeth so brilliantly put it in a post-*invisible Zürichs* Facebook conversation (21.05.2014) in which Sabeth, Nina, and myself were preparing for our next project, *stadtARCHIV*, while reflecting upon our previous experiences:

> What has so far and still is our problem: people don't get us. Other academics maybe, maybe other artists, sometimes – not always. But not the 'normal' people. We need to find much more approachable ways [. . .] in order to make them understand what we're doing and aiming for . . . [. . .] Now for example I'm in contact with this boxing group in Basel, and I really have troubles explaining to them what we do, in a way that they – the boxers – actually want to take part in it [. . .] We already manage well to adapt from theatre people to city researchers to engineering students . . . But we haven't found a way, yet, to adapt to the boxers. The boxers . . . hmmm . . . How to make a workshop for boxers?

The issues of authorship and transparency are here brought to the fore. If 'normal' people did not understand or relate to the purpose of our workshops, didn't we then, by claiming to represent them, 'put words in their mouths' solely according to our own interests and agendas? And didn't we then also create a specific portrayal of the urban inhabitant based on a pregiven interest that many of the participants, in fact, did not fully understand? How transparent about our goals and intentions could we be in our practice without risking a unidirectional process that merely reflected the pre-existing identities and assumed interests of the participants, rather than ensuring that these are constituted through the workshop process itself?

To address this contradiction between a presumably free aesthetic space and the social and institutional reality of art with all of its implicit exclusions, one must ask such ethical questions about how art 'should' operate. As Bishop (2006) argues, these questions come at a cost. Once the artistic practice is assessed in ethical terms, a limit is placed on the critical autonomy of the artist. As this part of the book will illustrate, we struggled with keeping our artistic aims of producing poetic and multilayered workshops that would resonate across many registers, and at the same time working not on grounds of our own choosing, but within various institutional constraints and expectations. While trying to keep our artistic aims, we were also weary of being perceived as elitist, and were thus, following the advice of Kester (2004), constantly on guard against our privilege and authority in order to create an egalitarian and open space. These considerations point to an important contradiction: If we already acknowledge the sociological and institutional critique of art that identifies it as an elitist locus of authority, how, then, can it simultaneously be an open space free of social and material constraints?

This part of the text will discuss this contradiction in light of a range of zURBS projects that succeeded *invisible Zürichs*. In these projects, we further developed the approaches we thought had worked and tried to redevelop the approaches we thought had failed in *invisible Zürichs*. This was a constant process of trial and error. With the project *stadtARCHIV* in Basel, we moved out of the exclusive space of the cultural institution and into the street in order to create an open space that was not polluted by the 'logic of the institution'. However, our quest for 'openness' led us to focus on harmonious reconciliation and 'togetherness' – an approach that risked stereotyping the participants as it did not sufficiently engage with their views and perspectives. The series of workshops we conducted as part of the project *Montopia* in Monthey was oriented around highly structured formats of collaboration that would ensure that the participants would engage with each other's views and opinions on a one-to-one basis. However, these formats ignored the potential for antagonistic encounters based on everyday prejudices relating to the participants' predefined social situations and identities in the city. They also ignored the antagonistic camps of 'us' and 'them' produced between those who were able to play and those who were not able to play along with the imaginative framework of *Montopia*. The *St. Clement's Utopolis* project in London sought to avoid such antagonistic encounters by concentrating on the ethical, rather than artistic, framework of the workshops. However, this approach failed to provide the participants with a transformative experience that would allow for new subjectivities to emerge and discoveries to be made by expanding the participants' capacity to act, engage, and connect.

The projects discussed in the next three chapters, then, illustrate the challenges posed by the inherent contradiction within socially engaged art in terms of aiming to cross the boundary between art and the social, and at the same time stand apart from the social as a distinct critical space. Hence, while the previous chapter could be concluded with a certain scepticism towards our initial aim to facilitate a so-called 'open process' in *invisible Zürichs*, this chapter illustrates how the idea of 'untaintedness' is heavily challenged by the particular frames of representation that are made available in socially engaged art. These challenges warn us against reducing the representative frames produced in this form of practice to any simplistic principle.

Bibliography

Bishop, C. 2004. "Antagonism and Relational Aesthetics." *October* 110: 51–79.

———. 2006. "The Social Turn: Collaboration and Its Discontent." *Artforum* 44 (6): 178–183.

———. 2012a. *Artificial Hells*. London: Verso.

———. 2012b. "Participation and Spectacle: Where Are We Now?" In *Living as Form: Socially Engaged Art from 1991–2011*, edited by N. Thompson, 34–45. New York: Creative Time.

Dryzek, J. S. 2000. *Deliberative Democracy and Beoynd*. Oxford: Oxford University Press.

Elden, S. 2004. *Understanding Henri Lefebvre*. London: Continuum.

Kester, G. 2004. *Conversation Pieces*. Berkeley, CA: University of California Press.
Purcell, M. 2014. "Possible Worlds: Henri Lefebvre and the Right to the City." *Journal of Urban Affairs* 36 (1): 141–154.
Saward, M. 2005. "Governance and the Transformation of Political Representation." In *Remaking Governance: Peoples, Politics and the Public Sphere*, edited by J. Newman, 179–196. Bristol: Policy Press.
———. 2008. "Authorisation and Authenticity: Representation and the Unelected." *The Journal of Political Philosophy* 17 (1): 1–22.
———. 2010. *The Representative Claim*. Oxford: Oxford University Press.

7 *stadtARCHIV*

The invited space and the logic of the institution

Pushing participation

invisible Zürichs brought zURBS attention and recognition as an artistic collective. Through our collaboration with the theatre Gessnerallee, which provided us with the residency for developing and executing the project, we got invited to make an alternative city archive, similar to the one in *invisible Zürichs*, at the theatre Kaserne in Basel, approximately one hour by train from Zurich. This archive, named *stadtARCHIV* (cityARCHIVE), was to be part of the performaCITY festival (6–15 June 2014) organized by the theatre in collaboration with the curator and researcher Imanuel Schipper. The festival was oriented around 'the issue of "City and Art"' (Trans4mator 2014) and featured performances and installations by well-known artists such as Dries Verhoeven, LIGNA, Ant Hampton, Tim Etchells, and Gob Squad.

Whereas these artists presented internationally renowned work that targeted a national and international performance audience, zURBS was cast as what we, inside zURBS, call a 'participatory alibi' that would ensure 'local anchorage' and participation. We were to engage various local community groups in a series of workshops similar to those we conducted in the context of *invisible Zürichs*. As in Zurich, the outcome of the workshops would be the creation of an alternative city archive that would (re)present lived experiences of Basel by letting the residents decide what aspects of the city should be archived. The *stadtARCHIV* was claimed to provide an 'authenticating frame' (Bell and Beswick 2014) that would offer the festivalgoers, of whom many came from abroad or from other cities, a direct insight into what the residents of Basel thought about their city. This authenticating frame tends to conflate authenticity with inclusion. As Nancy Thumim (2012) points out, contemporary art institutions need participating publics in order to establish and maintain their legitimacy as public institutions. Similarly, performaCITY needed the inhabitants of Basel to participate in the festival, in order to legitimate itself as a public festival for the 'ordinary' people of Basel in addition to the national and international festival crowd. Accordingly, zURBS was encouraged to recruit participants who would not normally attend the festival. This proved difficult, as we were not physically present in Basel prior to the festival, and thus did not have time for our meticulous netwalking efforts.

One week to go to the opening of the festival and we had still not succeeded in setting up workshops with local community groups. Frustrations grew, leading to a heated internal debate on the topic of participation. Sabeth felt uncomfortable with what she termed as 'pushing people to participate'. Although at the time I strongly disagreed (what was the point in what we were doing if we were neglecting the participatory element – the foundation of our work?), I do – in retrospect – partly understand her position. As was made clear by the episode with the couple who were interested in participating in the *invisible Zürichs* workshop but changed their mind the moment they had to enter the theatre, it does not matter how equitable the intentions that inform the creation of an arena for participation might be. Among people who enter a particular arena, their existing relationships to that arena will still influence their engagement with it. There is, as Sabeth might have argued, a stark contrast between spaces that are chosen, fashioned, and claimed by those at the margins, and spaces into which those who are considered marginal are invited.

The structural constraints of the cultural institution

The same day that zURBS was arriving in Basel, there was a big demonstration organized by the city's autonomous scene. The reason for the demonstration was that the Swiss squatter collective Wagenplatz was getting evicted by the police. Since May 2013, Wagenplatz had occupied a big vacant lot by the harbour. A cultural redevelopment had long been planned on the site, but it was only when the company Shift Mode won the bid to develop the site and decided to place a car park where Wagenplatz was set up that the police and city government finally got the opportunity to remove significant parts of the squatter community. During the eviction, the police acted overly aggressively towards the peaceful squatters, arresting 36 people.

We decided to invite the squatters to become part of *stadtARCHIV* by archiving remnants of the site in the alternative city archive, and this way give them the opportunity to represent their cause and even winning over supporters through our project. However, the squatters promptly declined our offer. Now, why did they decline?

Answering this question foregrounds the constitutive act of representation. Given the attention Wagenplatz had received due to the eviction and demonstration, the squatting collective had become a charged political phenomenon. By occupying the vacant lot in the harbour, Wagenplatz was already getting on with the work of representing itself. Carlos Frade (2011, cited in Purcell 2013) calls this a politics of 'presentation', people gathering to physically present themselves to the city government, as exemplified by the actions of 'the indignados' in Spain and the Occupy movement. This presentation demonstrates in physical space the gap between city representatives and those who are represented. While the city of Basel argued that it was acting in the best interest of the city, the Wagenplatz supporters insisted that they *are* the city, as much as any other residents. In this vacant space on the margins of the city, the collective had thus created a so-called 'site of radical possibility' (hooks 1989) from which they were able to define themselves in opposition to the conformity and hierarchical authority structures of landlords and the city government. The question, then, was what would happen to this political statement if it became part of an arts festival, and was thus perceived as art.

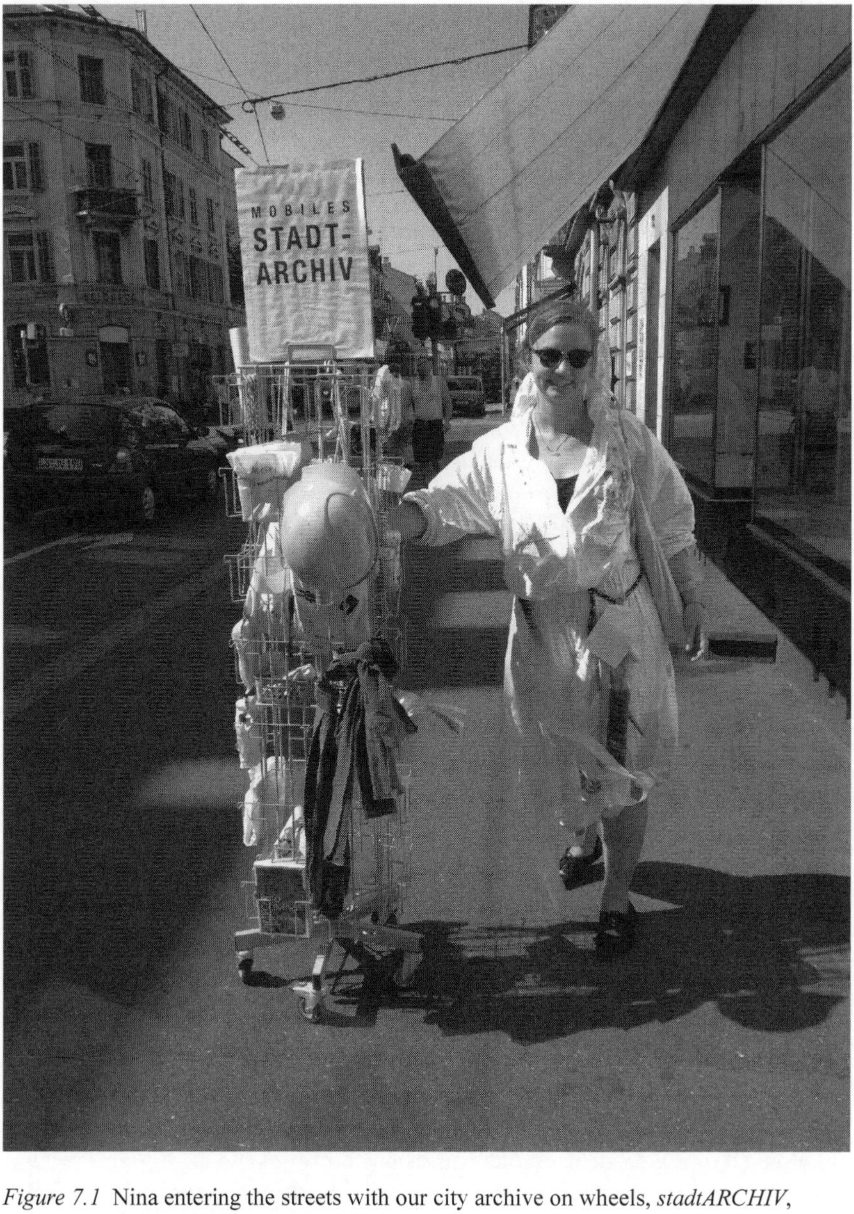

Figure 7.1 Nina entering the streets with our city archive on wheels, *stadtARCHIV*, zURBS, 2014 (image author's own).

The 'logic of the institution' is relevant here. Even though zURBS aimed at suspending this logic by creating an open space for the articulation of various identities and differences, *stadtARCHIV* was hosted by a state-sponsored cultural institution that produces art. This context bounded the possibilities for agency as well as inclusion for Wagenplatz because the only subject-position available here was that of an artwork. Accordingly, the collective risked being represented as a somewhat static (and lost) movement on display – a cultivated pictorial and poetic symbol of the protest rather than as a political criticism. Hence, by aiming to provide Wagenplatz with the possibility of using *stadtARCHIV* as their instrument for advertising/representing their cause and potentially even winning over supporters, we forgot that this instrument, the art institution, has its own set of rules that opposed the collective's goals. Claims to a so-called 'untainted' representation free of formal hierarchies and institutionalized structures are here challenged. The 'invited space' in which we wanted Wagenplatz to participate would intimately affect Wagenplatz' ability to enter and exercise its voice on the collective's own terms. The political field that Wagenplatz was part of hence defined the limits of how they could be represented in terms of the subject–object links that could be made in the given context.

Acknowledging these structural constraints of the 'invited space' of *stadtARCHIV* led to heated debates within the zURBS team. Sabeth used the example of Wagenplatz to argue that we had to stop being so occupied with participation and accept that some people might refuse to speak in the settings we created. What if, for once, we could use this project to experiment more for ourselves, developing methods for alternative urban research in the city, rather than chasing potential participants around town? I insisted on not giving up on participation. What was the point in what we were doing if it was no longer participatory? As it was all about to escalate, Nina came to the rescue: What about combining the two? What if we made events out in the city? This way, we could both test our research methods in the city and at the same time meet people where they are instead of trying to 'lure' them into the separate realm of the workshops. Sabeth and I were immediately keen on the idea. We decided to create a 'mobile research station' from a postcard holder on wheels. We filled the holder with archiving tools (glass jars, plastic bags, reagent tubes, papers, pens, scissors, chalk) and put a sign reading 'the mobile *stadtARCHIV*' on top of it. We agreed that we would tell the people we encountered that we were doing research on how to archive a city and ask them to help us archive Basel in that regard.

Moving into the streets

At the busy node Claraplatz, our first stop with 'the mobile *stadtARCHIV*', the Dutch artist Dries Verhoeven had installed his performance piece *Ceci n'est pas . . .*, which was also part of the performaCITY festival. Similar to *stadtARCHIV*, the piece aimed to trigger debates and discussions on Basel and its attitudes, what it chooses to put on pedestals and what to ignore. Both Verhoeven's and zURBS' projects aimed to engage a broader audience than would visit a theatre or museum, and thus articulate and make visible perspectives that, had they been located in art institutions, might not have appeared at all. However, there is a crucial difference in terms of how these perspectives were represented within the artistic practices themselves. This difference deserves some further scrutinizing as it relates to the inherent contradiction between aesthetics and ethics in socially engaged art.

As Bishop (2012) points out, artists' motivations for turning to social participation are often based on the same claim: 'contemporary capitalism produces passive subjects with very little agency or empowerment' (p. 1). Participation is a means to rehumanize a society rendered dumb and fragmented by the repressive instrumentality of capitalist production. Artistic practice, then, can no longer revolve around the construction of artworks to be consumed by passive bystanders. The art piece is put in the street for them to co-produce and not simply consume. This was very much the case for Verhoeven's performance piece *Ceci n'est pas ...*. According to Verhoeven (cited in CC's Lifestyle 2014), public space has lost its function as a place where people come together to discuss things that are important to society. People have become more and more indifferent, and it takes a provocation to make people stop, have a look at something, and reflect upon it. With *Ceci n'est pas ...*, Verhoeven wanted to bring the viewers in contact with their own thoughts, as well as the thoughts of other members of society with different opinions, and this way encourage a discussion. In order to do this, he had installed a glass box at Claraplatz. For 10 days, different scenes that we do not generally encounter in public space were displayed within the box: a woman with a genetic disease sitting in high heels on a bar stool, drinking cocktails, and flirting with passers-by (title: *Ceci n'est pas notre désir*/This is not our desire); an old naked woman (title: *Ceci n'est pas mon corps*/This is not my body); a man with a beard sitting on a Persian carpet and wearing a bulletproof vest (title: *Ceci n'est pas notre peur*/This is not our fear); a black man in schackles, wearing only a colourful necklace and a white cloth around his waist (title: *Ceci n'est pas de l'histoire*/This is not history); and so on.

Like René Magritte's famous painting *Ceci n'est pas une pipe*, the performance interrogates the 'fiction of representation' (language, agency, and politics in a constructed world) by disturbing the relation between words and images, reality and representation. The political potential of the piece, according to Verhoeven, is the discussion that is formed around the display case in which people are arrayed like rare relics (Karhunen 2014). The person(s) in the glass box cannot hear what the passers-by are saying, and thus the need for politeness and 'correct' speech abates. An appointed 'eavesdropper', in this case the theatre scholar Deborah Neininger, who is located among the audience, documents the opinions and views uttered by the passers-by, and puts them in a blog for the performance piece:

> Even deep-seated racial prejudice comes to the surface and erupts into defamatory comments. An elderly man walks past, thinking he is being funny and exclaims [to the black man in the glass box]: 'Do you want a banana?' A couple of bystanders laugh mischievously, but a young man is annoyed: 'You don't say things like that. This is a person!' 'More a monkey' another bystander comments. The word 'nigger' falls more than once in this hour.
>
> (Nenninger 2014)

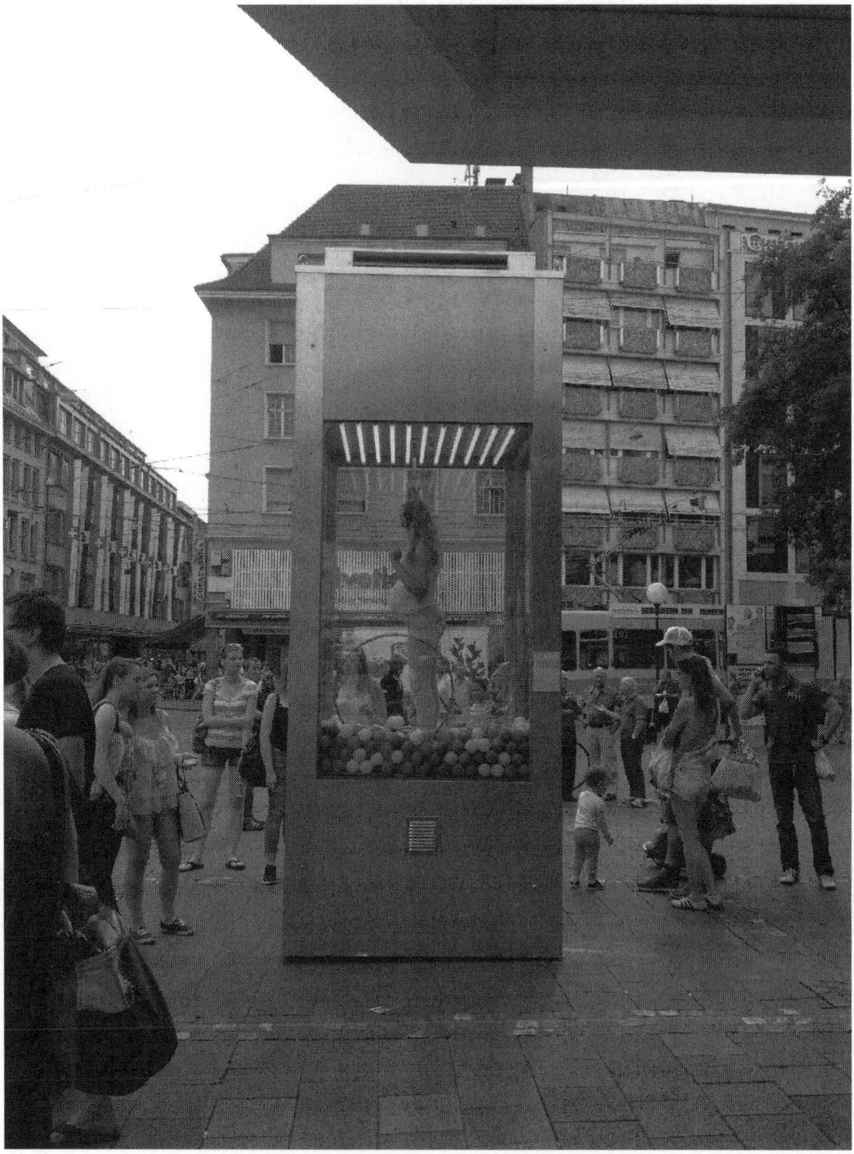

Figure 7.2 *Ceci n'est pas ...*, public performance in Basel, Dries Verhoeven, 2014 (image author's own).

Ceci n'est pas . . . was a provocation, aiming to produce antagonistic encounters between performers and audience, as well as between the audiences, in order to encourage them to question fixed identities and stereotypical images. For Bishop (2004), antagonism and confrontation are key elements of the political integrity of an artwork. I agree that confrontation has the potential to create encounters that may remind us of the illusory nature of coherence and show us that our perceptions and identities are shifting, unstable, and contingent. However, the fact that the antagonistic encounters produced by *Ceci n'est pas . . .* were being observed by an appointed 'eavesdropper' who published the reactions from the audiences in a blog without their knowing raises questions about the morality of this work. Bishop argues that these moral questions are irrelevant because they putatively interrupt artistic autonomy and the evaluation of the work's potential to articulate antagonism. But I am wary of excluding questions about morality on these grounds. The blog enacted a specific participatory identity for the participants by tending to divide the audience into philistines and cognoscenti. It conveys how a large part of the non-art public seems to not 'get it', asking questions such as 'What is the person in the box doing?' or 'How is this art?' or 'What does this have to do with Magritte, where is the pipe?' The only viewers who really seem capable of grasping the full complexity of the artwork are the educated art audience, such as the woman writing the blog. *Ceci n'est pas ...*, then, risks addressing exclusively those inside the enchanted circle of art, and therefore reinforcing a structural inequality (Charnley 2011).

This argument is at the core of Kester's critique of the antagonistic approach that he associates with a form of avant-gardism that is elitist because of its refusal to sufficiently engage with a non-art public. To Kester (2004), a collaborative artwork should not start with the a priori idea of an artist, but be a cumulative process of exchange and dialogue with the community within which the work will be produced. Accordingly, we – the zURBS team – wanted to create a process in which personal perspectives could be expressed on the initiative and terms of the participants, rather than as a reaction to some form of antagonistic provocation. However, foregrounding an ethical engagement with potential participants was easier thought than done. For example, at Claraplatz, we were approached by a middle-aged woman. She was curious about our *stadtARCHIV* installation and seemed very interested to talk. We told her about the archive and asked her if she wanted to share what Basel is to her. She immediately replied, 'Shit (scheisse), to me, Basel is shit'. We asked her why that is, and told her to write it down on the paper we had given her. She happily wrote on the paper, 'Basel is shit because of all the immigrants come and take my work. And after they arrived to the city, it is full of dirt'. In our attempt to accommodate marginalized or less socially accepted voices, such as this one, we failed to engage critically with these kinds of statements in ways that would attest to their relevance as a valid perspective on Basel. We were so occupied with making the woman who made the racist remark feel good about voicing her perspective, and in this way confirm the openness of *stadtARCHIV*, that we did not sufficiently question the experience, arguments, and ideas used to justify her view. As a result, the woman's remark was not given any further explanation, for example in terms of concrete examples from her personal experience of Basel. Hence, the remark could easily be written off as simple-minded racism that has little or no relevance for the city as such.

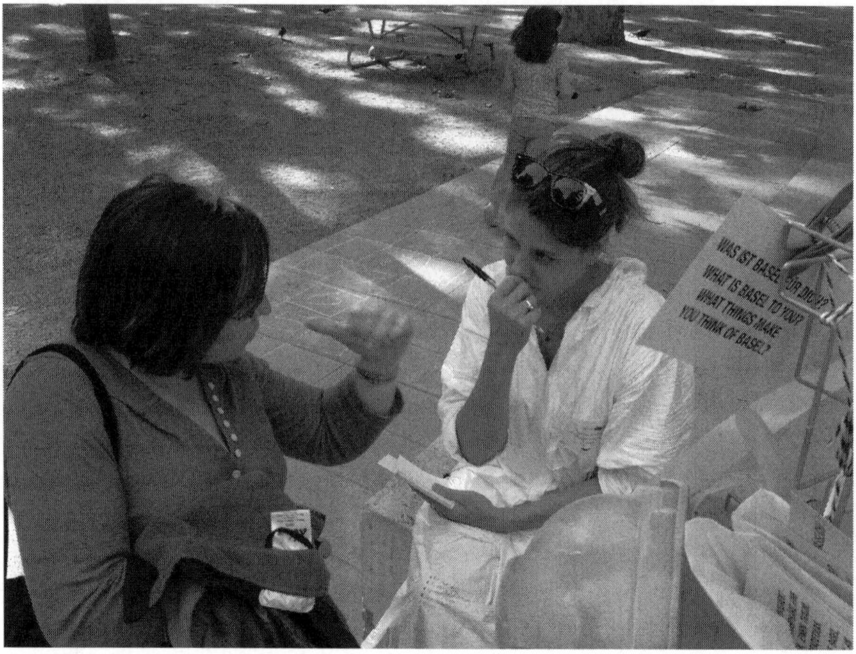

Figure 7.3 Talking with people in the street, *stadtARCHIV*, zURBS, 2014 (image author's own).

Bishop's antagonistic approach could have provided more polemical grounds for dealing with this kind of statement. Rather than achieving a harmonious reconciliation focusing on an uncritical 'togetherness', this approach could have sustained a tension between our different positions, and thus made explicit what is repressed in sustaining the semblance of social harmony. By avoiding this tension, we undermined the statement as a banal perspective, rather than validating it by engaging with it on its own terms. In this context, then, Kester's dialogical aesthetic can be criticized for falling short of addressing the political aspect of communication. This political aspect relies on recognizing a model of subjectivity that is not harmonious, but divided into partial identifications open to constant flux: 'the presence of what is not me renders my identity precarious and vulnerable, and the threat that the other represents transforms my own sense of self into something questionable' (Bishop 2006, 66).

By seeking to sustain social harmony rather than disrupting it, we repeatedly failed to engage substantially with the various perspectives on Basel that we were presented with. We did not ask the man eating lunch in the park why 'rain, simply rain' was his immediate description of Basel, or the man waiting for the tram why he insisted that Basel was no different from any other city, or why the function of the river to gather all inhabitants on the same side of it was so important for the architects. As a result, we did not feel comfortable with the intervention as it seemed only to reflect certain unnuanced ideas about Basel rather than challenging them. We agreed that we would need more time to properly develop a concept that would enable us to engage with passers-by in ways that did not override critical perspectives, and which at the same time acknowledged the tension between our artistic framework and the non-art public. As we did not have that time in Basel, we decided instead to focus on the workshops that we had committed to conduct in the frame of the festival. Acknowledging that we could not ignore the 'taintedness' of our practice in terms of its structural constraints as part of this specific art context, we now focused on the connections and collaborations that could be established within *stadtARCHIV*, between zURBS as facilitators of the artistic and participatory process and the creators of and visitors to the archive: How could we promote an awareness of difference in the workshops and at the same time build commonalities among the participants? How could we acknowledge tensions between different views and still facilitate an openness where everyone would be encouraged to have their say?

Practising artistic authorship and curating the archive

The archive was located at the festival centre where all the festivalgoers mingled in between the different events. The festival organizers emphasized that it should be possible to understand and interact with the archive without any (prior) knowledge of the workshops. The chaos and deliberate lack of structure of the *invisible Zürichs* archive thus had to be 'tamed' in favour of more coherent and communicable narratives so that the archive could be more easily read by 'outsiders'. Accordingly, we – the zURBS team – completely appropriated the process of archiving in the workshops. The participants still provided the objects and the stories, but we took care of labelling and placing the objects in the archive. Nina, with her eloquent handwriting, wrote all the labels as short, poetic summaries of the discussions connected to each object. We then grouped the objects into categories relating to the different tasks the participants of the workshops had carried out in order to collect them, so as to ensure connections between the objects or between the objects and the ideas connected to them. The note from the woman ranting about immigrants taking her work was, for example, put in the category 'Traces of conflict' together with broken pieces of mirror, broken records and police cordons from the Wagenplatz site, pictures of street graffiti, and plastic straps that one participant explained are often used by the police as substitutes for handcuffs.

The Basel archive marked a significant shift in the mode of representation from the *invisible Zürichs* archive. The Zurich archive provided an overwhelming and all-encompassing sphere of unstable cardboard boxes, shrinking and bending under the weight of the objects that were piled inside and on top of them, stacks of photographs randomly taped to the walls, fragmented and often unreadable labels due to various handwritings, and heaps of garbage lying around on the floor.

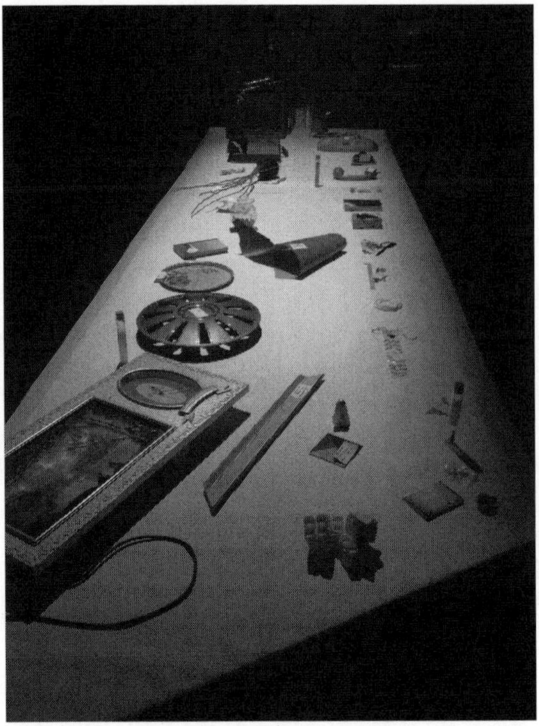

Figure 7.4 Exhibition display, *stadtARCHIV*, zURBS, 2014 (image author's own).

The Basel archive, on the other hand, was oriented around the display of singular objects lit by spots, placed on shelves covered in bright yellow glossy paper, neatly separated so that each object could be scrutinized as an independent entity, much like paintings in a museum. Although the Basel archive was easier to 'read', it also enforced an approach that fostered distance rather than proximity to the archive. This distance can be seen as directly linked to its unified aesthetic, which suggested zURBS' particular artistic authorship and voice, with the power to conserve and interpret the archive's material by highlighting particular topics and features as worth attention and selecting stories to tell. As a result, the archive was often referred to as zURBS' archive, and not the archive of Basel or its residents. The visitors accordingly treated the archive very much like an exhibition they simply walked through from one end to the other, rarely stopping for a closer look or for any form of contemplative immersion.

This was a stark contrast to the *invisible Zürichs* archive, where many visitors would spend much time immersing themselves in the sprawling and overwhelming complexity of the material. Similar to Sigman's huts, the DIY aesthetics of this archive suggested a non-hierarchical mode of representation where the material triggered personal experiences, observations, and questions to be shared among the visitors of the archive. As Paula, a cultural worker in her 30s, explained to me, while sitting comfortably on some pillows in the Zurich archive sipping a cup of tea:

> This is about the invisibility, like people have experienced something that I do not know, but then I see the things written and then I want to know the story. It is a little bit like this game, that you have cards and a person tells about his card and puts it down in the stack and then you have to guess which card he was talking about. This is a little bit like that. Like you are trying to catch experiences that are invisible to you.

The Swiss artist Thomas Hirschhorn points to a process of 'implicating the audience' where visitors are confronted and forced to engage with the material that surrounds them. This is done by creating a form of 'spatial montage', or, as Hirschhorn himself calls it, 'a mental space, to envelop everything' (cited in Riding 2004). In Hirschhorn's work, all elements are 'juxtaposed with myriad others, creating a swarm of contrasts and unexpected correspondences between different representations' (Gardner 2012, 38). Within this space, the actions and momentary interactions between visitors are inextricable from the aesthetic sphere enveloping and contextualizing them; they too are juxtaposed or correlated with the surrounding sprawl of imagery, text, and other forms of discourse. As the art student Mirjam Wirz observed when making her artwork based on the *invisible Zürichs* archive, the words appearing in the texts in the archive were repeated in the conversations appearing in the room, emphasizing how the material provided by the workshop participants and the dialogues between the visitors and participants were all one unity.

However, many workshop participants as well as visitors to the *invisible Zürichs* archive found it chaotic and overwhelming, and often had trouble navigating in it. As one of the participants, an older female university professor, pointed out:

> For me, the archive wasn't that interesting when I came here and saw it. Because we did it [the workshop] we [now] know the experiences behind the objects. But for people who did not experience this, it is pretty hard to read. Because it is quite chaotic and 'dumb'. And I think that is where you [the facilitators of the workshops] come in, you have to make these things easier to read.

The modes of representation in the Basel archive and the *invisible Zürichs* archive had their advantages and disadvantages in terms of representing various identities and interests, and on the basis of these representations build commonalities and awareness of differences regarding how we imagine, desire, and understand cities. The *invisible Zürichs* archive may have succeeded in establishing connections between the archive and its visitors – participants of the workshops as well as other visitors. However, its complexity compromised the readability of much of its material, and thus may have alienated people who did not understand the role and function of the archive. The Basel archive, on the other hand, may have made the material in the archive easy to read, but as a result compromised people's sense of ownership and engagement by giving it a specific artistic authorship.

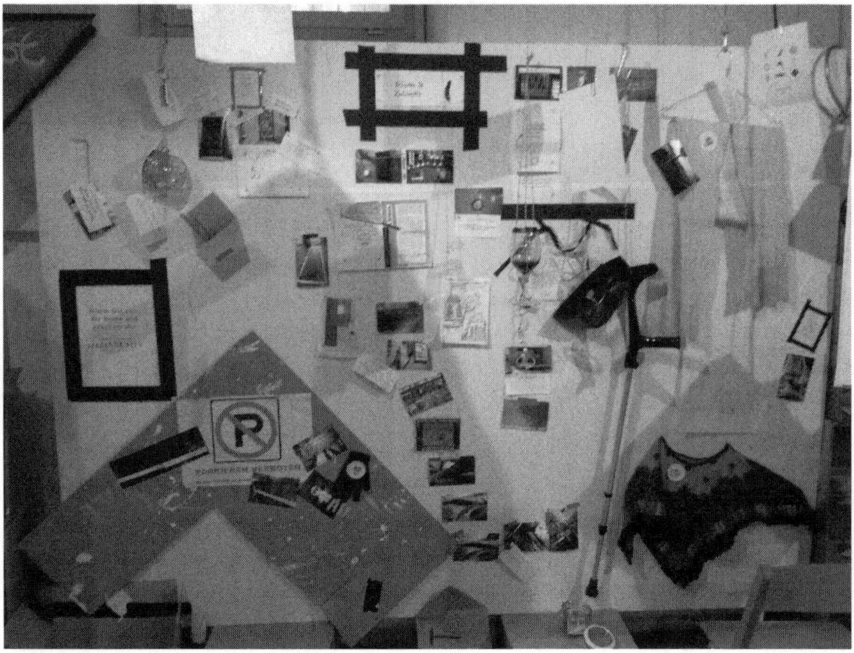

Figure 7.5 Exhibition display, *stadtARCHIV*, zURBS, 2014 (image author's own).

Conclusions

As this chapter illustrates, zURBS' aim to produce a so-called 'open' and 'untainted' space that would be disconnected from formal hierarchies and institutionalized structures was heavily challenged by the particular frames of representation that were made available in our *stadtARCHIV* project. We became painfully aware of the 'cultural logic of the institution' when trying to recruit the squatters of Wagenplatz to represent themselves in the archive. And even though we moved the *stadtARCHIV* into the streets in order to escape this logic, we did not manage to change the cultural field that defines the limits of the representable. While being wary of not alienating passers-by and enforcing a distinction between those who are inside and those who are outside of the art discourse, we focused so much on openness and agreement that we risked stereotyping the participants by not sufficiently engaging with their views and opinions. In the archiving process of the workshops, on the other hand, we overly engaged with the contributions of the participants. In our quest to make the archive 'readable' and understood by audiences that had not been part of the workshop process, we took the roles as main archivists in order to ensure direct connections to be made between the participatory process and its representation. However, the resulting representational framework of the archive became a form of gallery/art exhibition that compromised the participants' sense of ownership of the archive.

The questions, then, were how to create a middle way between these two somewhat hierarchical and anti-hierarchical representational approaches. How, amid the cacophony of interests, identities, values, desires, wants, and judgements, to create a common ground and collective ownership where people can address one another as co-producers of the city. How to include perspectives anti-hierarchically and still ensure a clearly communicable and understandable outcome. With our next project, *Montopia*, we tried to address these questions.

Bibliography

Bell, C. and Beswick, K. 2014. "Authenticity and Representation: Council Estate Plays at the Royal Court." *New Theatre Quarterly* 30 (2): 120–135.

Bishop, C. 2004. "Antagonism and Relational Aesthetics." *October* 110: 51–79.

———. 2006. "The Social Turn: Collaboration and Its Discontent." *Artforum* 44 (6): 178–183.

———. 2012. "Participation and Spectacle: Where Are We Now?" In *Living as Form: Socially Engaged Art from 1991–2011*, edited by N. Thompson, 34–45. New York: Creative Time.

CC's Lifestyle. 2014. *Dries Verhoeven – Ceci n'est Pas ….* Available at: www.ccslifestyle.com/2014/06/dries-verhoeven-ceci-nest-pas.html [accessed 10.06.2016].

Charnley, K. 2011. "Dissensus and the Politics of Collaborative Practice." *Art & the Public Sphere* 1 (1): 37–53.

Frade, C. 2011. *Events in Spain. The Rise of the Indignant: Spain, Greece, Europe*. Available at: http://backdoorbroadcasting.net/2011/06/the-rise-of-the-indignant-spain-greece-europe/ [accessed 10.06.2016].

Gardner, A. 2012. "De-Idealizing Democracy: On Thomas Hirschhorn's Postsocialist Projects." *Art Margins* 1 (1): 29–61.

hooks, b. 1989 *Yearnings: Race, Gender and Cultural Politics*. New York: Routledge.

Karhunen, M. 2014. *Interview with Dries Verhoeven*. Available at: http://balticcircle.fi/2014/en/events/dries-verhoeven-ned-ceci-nest-pas [accessed 11.10.2018].

Kester, G. 2004. *Conversation Pieces*. Berkeley, CA: University of California Press.

Nenninger, D. 2014. *Dries Verhoeven 'Ceci n'est Pas'*. Available at: http://cecinestpasbasel.tumblr.com [accessed 10.06.2016].

Purcell, M. 2013. *The Down-Deep Delight of Democracy*. Chichester: John Wiley & Sons.

Riding, A. 2004. "Dissecting Democracy, Swiss Artist Stirs Debate." *The New York Times*, 27 December. Available at: www.nytimes.com/2004/12/27/arts/design/dissecting-democracy-swiss-artist-stirs-debate.html?_r=2 [accessed 10.06.2016].

Thumim, N. 2012. *Self-Representation and Digital Culture*. London: Palgrave Macmillan.

Trans4mator. 2014. *About Performacity*. Available at: http://performacity.net/?page_id=746 [accessed 10.06.2016].

8 *Montopia*

Collaborative inventions and antagonistic encounters

From archive-making to model-making

invisible Zürichs did not only catch the attention of the performaCITY festival in Basel. In January 2014, a few weeks after the closing party of *invisible Zürichs*, we were contacted by the festival organizer of Hik et Nunk – 'new festival of contemporary cultures' – in Monthey, inviting us to work with the local population 'to "redesign" the city in an artistic/utopian way' (invitation email from Hik et Nunk organizer, 08.01.2014). Particularly important for the organizers in this regard was that our project would have a clear end result in the form of an 'exhibitable' model that would 'sketch visionary or utopian urban projects' (personal email correspondence, 03.02.2014). A model-based project would, they argued, be 'simple to explain' and 'really interest the media and decision makers, particularly if we work with different kinds of users'. We – the zURBS team – accordingly proposed to make a model-making project aimed at opening up a debate concerning the future visions for Monthey, working with the notion of utopia as a critical model for anticipatory consciousness and open-ended expressions of desire for a better way of living (Pinder 2002). The idea of changing the format of our workshops – from the abstract archiving of invisibilities to the more concrete model-making expressing the participant's dreams and desires for Monthey – seemed like an opportunity to get at some of the challenges of *invisible Zürichs* and *stadtARCHIV*. Would this project help us find a middle way between being abstract and open and being unidirectional and leading when representing the wishes, desires, wants, and dreams of the participants? What forms of claim-making would this format enable on behalf of its participants?

On Saturday 30 August 2014, the zURBS team arrived in Monthey, a city of approximately 18,000 inhabitants located in the Swiss Alps, in order to find out. Ahead of us were five intense days of workshops with local groups. The festival organization had contacted and recruited the participants for us, and had done a great job in doing so. Drawing on their institutional network, they had lined up a broad range of participants, such as employees of the Crochetan theatre, members of an association for assisting people with psychiatric problems, politicians from the city government, architects from a local architect office, a school class of 16-year-olds, a school class of 10-year-olds, members of an organization for parents with schoolchildren, and a group consisting of various Monthey residents. Not having to spend so much time on getting participants had enabled us to spend more time on developing the workshops, playing around with the concept, as well as testing out and experimenting with different approaches, exercises, and formats.

The format we ended up with was the following: the participants were to create a model of *Montopia* – the Monthey of their dreams and desires. As the participants commented:

> I think the name is great obviously: Monthey and Utopia mashed together as a theme in a way [. . .] As a trigger I thought it was really great to think about what could this city be, what is *Montopia*, what is this virtual city that is hovering over Monthey and could take form one day.
>
> (Lucas, 40s, architect, personal interview)

Montopia was to be built out of objects that the participants would find in the city: a fountain made out of a snail shell and a green glass bottle, a beer can tree, a climbing gym made out of a mushroom, and so on.

The journey into the city to collect these objects was framed by a performative ritual where the participants were blindfolded and led into an elevator in which a soundtrack was played inviting them to enter Monthey as if it already was *Montopia*. The participants were then led, blindfolded, from the elevator and into the city, before they could remove the blindfold and start looking for relevant material. As one of the participants explained the experience:

> I think the ambience in the elevator is great to lose your habits, so you are entering in a mood. You have to lose control, it is something different. So, it is very clear that something different is happening. You are blind, so you are losing what you are used to seeing.
>
> (Maud, 40s, theatre administrator)

Back at the workshop location, the process of constructing *Montopia* out of the found objects (whether in the form of a physical place, building, means of transport, atmosphere, or the like) was facilitated as a form of speed dating: in pairs, opposite each other, the participants would sit with two collected objects in between them. They would carry out a 'task' relating to the objects. These tasks would, for example, start with having the participants tell and write down memories they have from Monthey that relate to the object, or, for example, finding five characteristics/qualities the objects have in common. It would then develop into building something with the two objects that they would like to have in *Montopia*. Between each 'task', the participants would move one seat to the right, whereas the objects stayed put. This meant that for each task, the participants would have to collaborate with a new person and a new object, taking into consideration the previous tasks relating to these objects, as documented by the other participants.

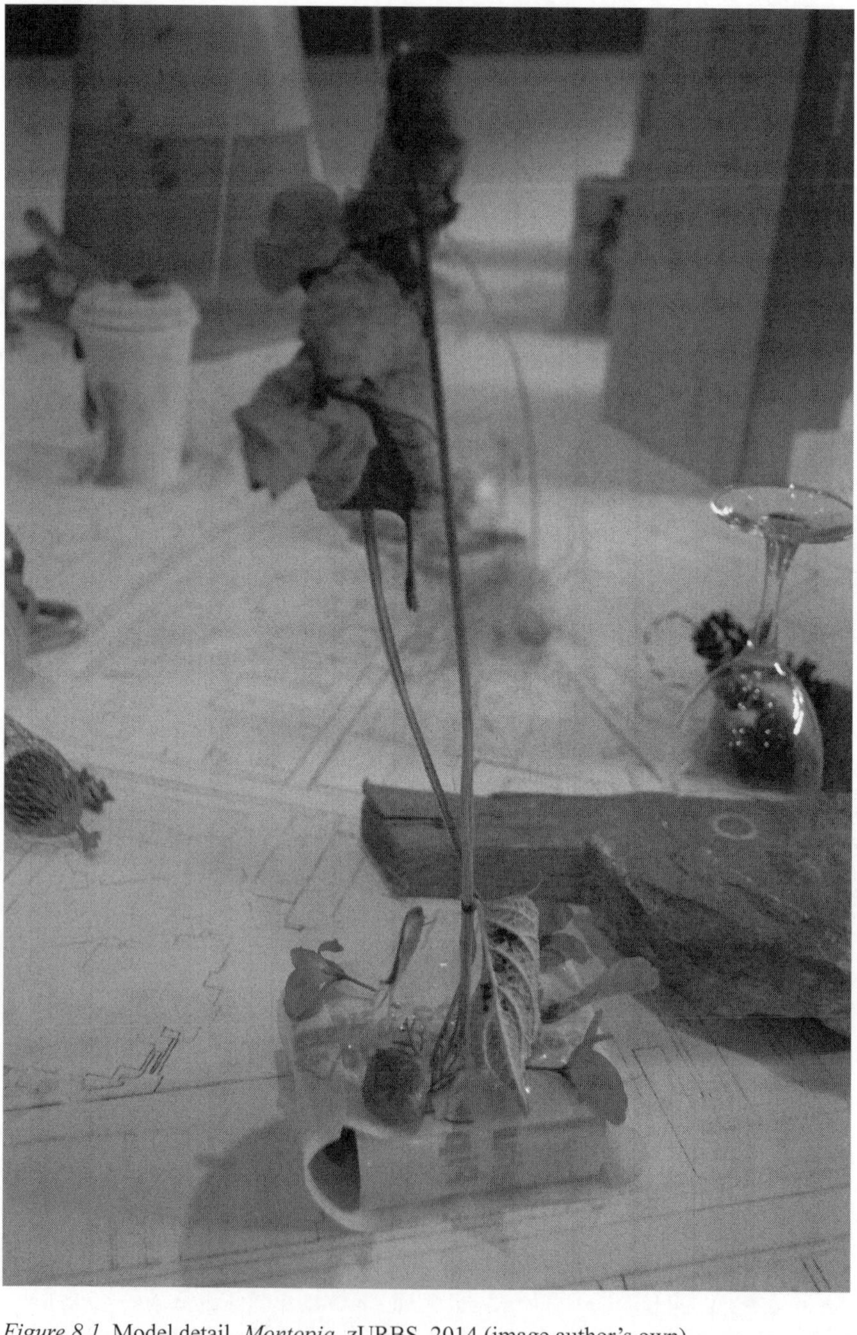

Figure 8.1 Model detail, *Montopia*, zURBS, 2014 (image author's own).

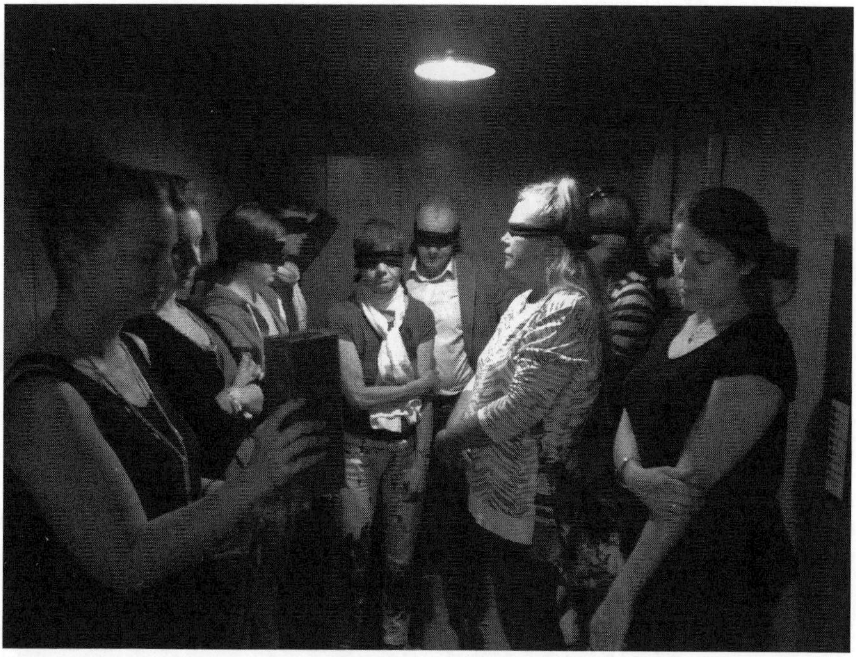

Figure 8.2 Performative ritual in the elevator, *Montopia*, zURBS, 2014 (image author's own).

Through this process, the participants obtained a shared ownership over all the constructions that eventually took shape. As one of the participants put it:

> It was very good because in the beginning you are with somebody and you think with this somebody, but then after two minutes you are with somebody else. So it is very strange because you do not speak the same way with the other person, you do not think in the same way, so you get a different perspective. And the third, and the fourth, and so on . . . So it is very rich, because you do not build something with only one person, or you are not just one person in a group, but you are one person with another and another and another.
>
> (Marco, 40s, theatre director, personal interview)

After the workshops, zURBS would go to all the places where the models had been placed on the map, take a picture of the place, and then Photoshop the model into the actual place where the residents had placed it. This way, a plausible reality was created out of people's dreams and imaginations. The model together with these photoshopped montages, as well as descriptions and explanations for each building, would then be exhibited.

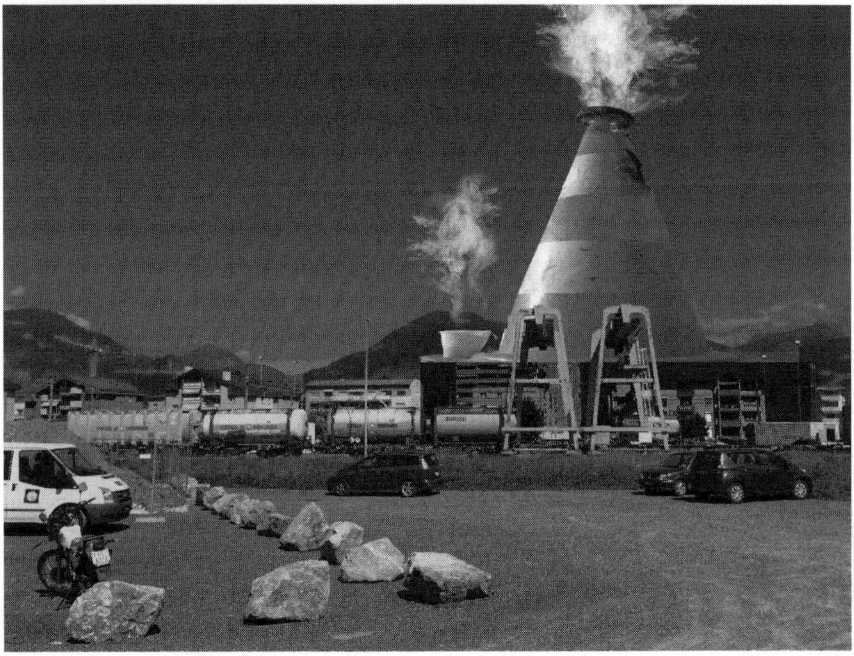

Figure 8.3 'The Cloud Factory', projection of Montopia in Monthey, *Montopia*, zURBS, 2014 (image author's own).

Utopia as a tool for imaginative visions

It was important for us to keep the aspect of collecting objects in the city from our previous projects. The objects ensured a direct link to the present city of Monthey and pointed to the transformative potential of both the objects and the city as such. The city of *Montopia*, then, was to be seen not as some kind of fantasy or the perfect city completely detached from reality, and thus as an escape from Monthey. Rather, the idea of *Montopia* was to enable a form of 'transduction' that would, as explained in Chapter 1, cut a path that leads beyond the actual world already realized and towards a possible world yet to come. This possible world does not exist outside or beyond our current situation; it is already here – immanent, incipient, coming. By inviting the participants to enter Monthey as if it was *Montopia*, we wanted to make this clear.

Similar to the framework of the *Invisible Cities* stories that we applied in the previous projects, *Montopia* provided a model for creativity and alternative positions from which to experience and represent urban space that adopted imaginative, transgressive, and poetic constructs. For example, one participant, who assisted the workshop with the 10-year-olds, told about how the kids, in the process of looking for objects that they liked and desired for building *Montopia*, became aware of all the garbage lying around in the streets. The kids agreed that they would not collect the garbage as material for *Montopia* as they did not want to have a city full of garbage. Instead, they put all the garbage they found in the designated trash cans. By attending to Monthey in new ways, as if it was the city of their desires, the kids realized that littering was not something they desired in their city. The model they would build of *Montopia*, then, was not a philosophical exercise in daydreaming, but a practical tool and powerful lens through which the participants could view their city:

> You had this link with what you found because it already has some history. Because you found it and you took it. So it is a history to link with the city clearly. And you had to reinvent it, because it had a function in the city and now you can invent another function for *Montopia* [. . .] You are reinventing something that already exists, but you just invent something else.
>
> (Maud, 40s, theatre administrator, personal interview)

> If we think about a city, in fact everything is possible. We can do anything in the city in fact, and the workshop proves it. So, even if the objects are not realistic, it shows that you can have imagination for the city. And that you can have great projects even for a small city. And this is very interesting. It is not because you are a small city that you have to have small ideas. And this is, for me, the main point of this workshop now in Monthey. Because I think there is a little bit of smallness in the minds here [laughs] and people don't . . . are not very courageous for the town. And it is not a problem of money, but it is a problem of openness and creativity.
>
> (Marco, 40s, theatre director, personal interview)

However, in focusing on particular visions and desires for Monthey, there was a risk for the workshops solely to echo a 'longing for singular pictures of the future or for supposedly stable and adequate representations of a "good society"' (Pinder 2002, 238). We were, for example, surprised to find that the majority of objects that were collected in the workshops were taken from nature, be it leaves, branches of trees, various flowers, snail shells, stones, and pieces of wood. When asking about this focus on nature as part of the participants' visions for Monthey, the answers pointed to the lack of green space and parks in the city centre. Due to its location by the river, Monthey was an industrial city hosting large chemical and steel companies such as Ciba-Geigy (now Novartis, Syngenta, and Cimo) and Giovanola. As the city was surrounded by nature, very few green squares and parks were constructed in the city centre. When the city grew, the distance to nature and green spaces increased, without any initiatives taken to provide the centre with any parks or the like.

The broad consensus regarding the need for more green spaces in Monthey could easily be cast off as representing a coherent and totalizing vision for the city. However, we realized that for some participants, this vision related to a desire for green 'breathing pockets' that would promote health and happiness, while for others it related to a desire for a more 'urban' vibe for the city. As one of the local architects explained, to have a public park in which you could have lunch or simply sit and read a book would be a sign of urbanity (personal interview, 05.09.2014) – a particular atmosphere and self-understanding that many residents thought Monthey lacks. *Montopia* aimed at representing these different perspectives and visions, not in totalizing and singular forms projecting representations of a 'perfect' city, but as attempts to address what is possible through collaboration, negotiation, and dialogue. The challenge, then, was how to stimulate an openness and attention to differences, while at the same time trying to represent collective visions for a city of our dreams and desires.

Speed dating, tall tales, and collective creations

> It is like if we became children, you know? And this is very good, because we do not play in life. Well, I play a little bit because I work in the theatre, but other people are not playing and are not very creative. And, you know, to be at the table with pens, it is like we were children. And it washes [off] all our social conditions and everything. So it is very refreshing.
>
> (Marco, 40s, theatre director, personal interview)

Several of the participants mentioned how the workshops made them feel like children playing games. The miniature model of *Montopia* had a childlike aesthetic, with its fountain made out of a green glass bottle, its beer can tree, its climbing gym made out of a mushroom, its cone temple, its glass tower of witches made out of an apple and a brown glass bottle, its interactive opinion sculpture made out of a takeaway coffee cup, its media library ('mediatheque') made out of plastic bottles taped together with wood and flowers, its hotel made out of an orange and white traffic cone, its energy station made out of a pine cone, its flying public transport made out of a brown leaf and a taped-together piece of wood, and so on. Parallels can here be drawn between the objects collected in the city and the toy. As Stewart (1993) observes:

> The toy is the physical embodiment of fiction: it is a device for fantasy, a point of beginning for narrative. The toy opens an interior world, lending itself to fantasy and privacy in a way that the abstract space, the playground of social play does not. To toy with something is to manipulate it, to try it out within sets of contexts, none of which is determinative.
>
> (p. 56)

The speed dating and tasks relating to the objects were attempts to make the participants 'toy' with the objects in a similar way, making them create imaginary conversations between different objects and imagine them to be buildings, means of transport, and so on. Links can here be drawn to the folklore tradition of the tall tale, in which a session starts with understatement and proceeds with each

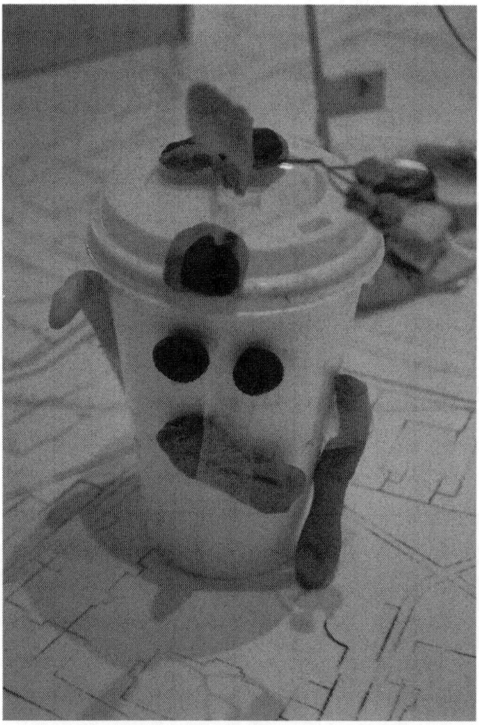

Figure 8.4 'The Opinion Sculpture', object made by participants out of a takeaway coffee cup, *Montopia*, zURBS, 2014 (image author's own).

narrative element to move further away from reality as defined by everyday lived experience. The tall tale recounts or invents experiences that are possible only in a fictive universe. Similarly, the speed dating was a process of invention through progressive stages. The top of a green glass bottle that was broken was, for example, initially collected for its beautiful material and sensory qualities: despite the broken glass, the green circle around the bottle neck was still intact. The glass had lost its initial shine, but was now beautifully blurry and semi-translucent, according to its finder. In the speed dating process, the bottle was seen as a reflection of Monthey due to its harmony (the circle), but also its sharp character (the broken glass). It was discussed as a symbol of green space due to its green and organic form, relating to the apparent lack of green spaces in Monthey. Later in the process, it came to represent the dangerous conflict between people and nature (sharp glass), and subsequently the privatization of water (due to its ability to contain water). Finally, all these understandings of the bottle top crystallized in the making of an arts pavilion for *Montopia*. A pavilion that would function as a green, open, and chaotic space for a form of contemplation that would help people feel alive. The pavilion was to be placed on top of a tall building so that it would be

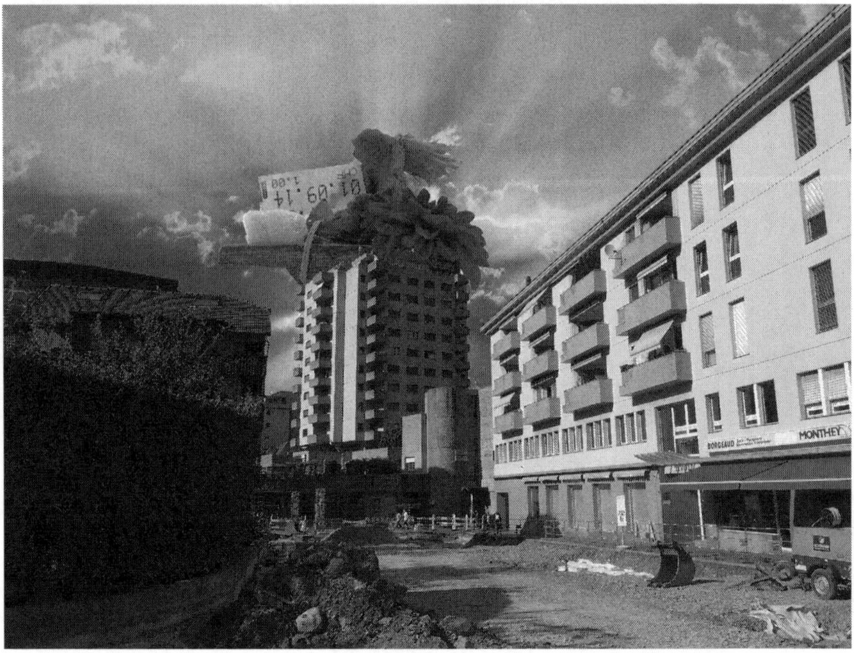

Figure 8.5 'The Arts Pavilion', projection of Montopia in Monthey, *Montopia*, zURBS, 2014 (image author's own).

protected from traffic. It was to be of no other use than simply being beautiful to look at; it would be turning around so that it could be seen from all angles.

Stewart (1993, 98) points to how tall tales are tied into a contextual structure from which they cannot be detached without a considerable loss of effect. The story about 'the one [fish] that got away', for example, 'is all the more credible because we only have the narrator as witness, yet all the more incredible because it is beyond the range of audience's experience'. The context of the tall tales, then, is that of an 'outside' position, in the sense that they are based on an often-solitary experience (e.g. the fishing trip), removed from the domestic and domesticated modes of sociability. *Montopia* provided a similar form of contextual structure, in that it was to be a city of imagination, removed from the constraints and limits of Monthey. As one of the participants put it:

> The people working in urbanism are really more like me than like you. And I think it is good sometimes to go outside these borders, outside what you do always. And I think it is not a bad idea to have a new vision of a city, a country or a place, or a street. It is sometimes good to go outside and see how you can look at this with other eyes.
>
> (Philippe, 50s, bank manager/urban planner, personal interview)

The stories told about *Montopia*, then, could take on fantastic features and fabulous events, such as, for example, 'The Glass Tower of Witches', which takes positive energy from above and distributes it around the city, deleting the arrogance that exists in people's heads and making wishes come true. However, despite their fantastic qualities, the stories expressed social and political criticism relating to present-day Monthey. The 'Tower' was made as homage to a glass factory that used to be located in Monthey in the 1980s. The participants, belonging to the organization for people with psychological problems, emphasized the openness of the factory and the many workplaces it had created. Hence, the homage can be seen as reflecting nostalgia over manual workspaces being lost and the desire for more workplaces that would suit the participants' needs.

These desires were not necessarily fully worked out prior to the workshops but were formed through the workshop process. As in the context of the tall tale session, which works cumulatively (each story sets a plateau for the following story to take as a basis for the possible), the desires and wishes expressed in the workshops often progressively moved from personal reflections on something missing from current conditions to statements and manifestations used as a base for expressing and articulating demands. In the workshop for the group of teenagers, for example, one of the boys at first seemed completely disinterested in taking part. He repeatedly stated that he 'did not have any opinions'. When time came to

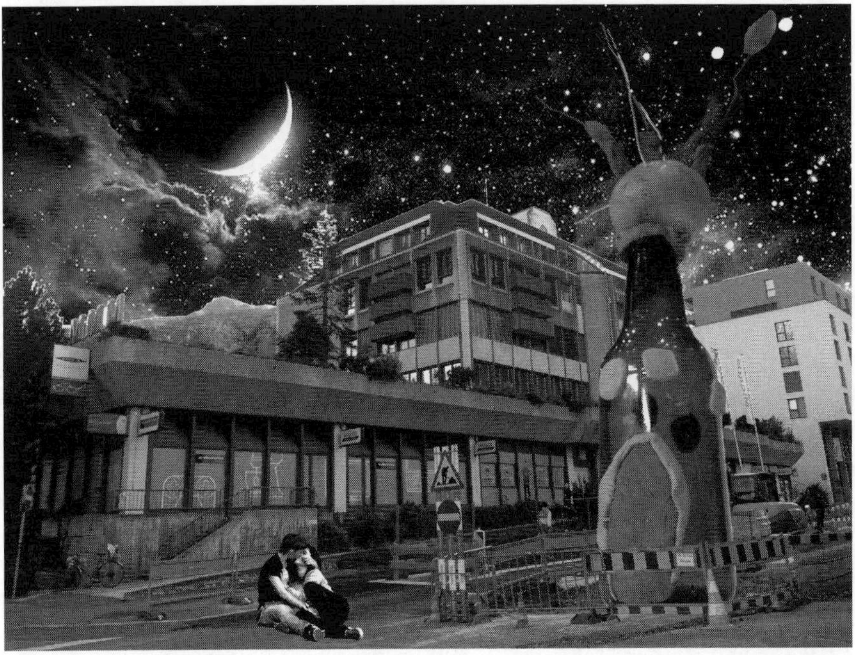

Figure 8.6 'The Glass Tower of Witches', projection of *Montopia* in Monthey, *Montopia*, zURBS, 2014 (image author's own).

make the models, however, he livened up and made a great effort to make a statue out of an empty Coke bottle. At the end of the workshop, when we were discussing each model, I asked him about the model. He jokingly said that this was a sculpture of Bob Marley – a weed smoker – and then laughed, showing off his indifferent attitude to his schoolmates. When I asked him why it was important for him to have a statue of Bob Marley in *Montopia*, he first raised his shoulders as if to say that he did not have any opinions on that matter. However, as I encouraged a reply, he slowly started to talk about how *Montopia* needs music and places for youth to hang out. After encouragement from his classmates, who applauded his suggestion, he explained that there are not many 'free spaces' in Monthey where he and his friends could play music and hang out without being told to go elsewhere by authorities. It was as if the boy realized, as he was speaking and engaging with the map and the model, that he actually did have an opinion about his city and a desire to change it. The 'Bob Marley sculpture', then, was turned into a manifestation of a youth culture demanding its right to the city.

Monthey versus *Montopia* and 'us versus them'

The risk with the stories of *Montopia* invented by the participants was that they could easily be cast off as mere exaggerations that had nothing to do with 'reality'. In one workshop, where the majority of the participants were employees at the local theatre, a central figure of the planning department in Monthey, Philippe, showed up. It was clear that Philippe created a dichotomy between the idea of utopia in the workshop and 'reality':

> Reality is not to have – how do you say that: noisette? – a nut is not speaking with the flower and that is not concrete for me [. . .] My second job is director of the bank here, and I am always on something really, really real [. . .] For me it is a little bit too much utopia. For me. Because I am so, I need to touch something, I cannot do with virtual things, I am so, that is my character [. . .] It is so special . . . I can't have the vision of the castle made out of an egg carton or something like this, it is not possible for me.

This perspective places the loci of authority in domains such as the bank – a place that deals with the 'real' world. To uphold this perspective, it is necessary that fantasy, exaggeration, and fictiveness are socially placed within the domains of anti- and non-authority: the childish, for example (Stewart 1993). In creating this dichotomy between utopia and reality, Philippe resorted to the technocratic approach that, as mentioned in the Introduction, resorts to arguments outside argumentation – a city is a city – and so decreed in advance which approaches to urban space are legitimate. By constantly referring to the real Monthey versus the unreal *Montopia*, Philippe treated the image of the city (*Montopia*) as a mere reflection or distortion of the 'real' city (Monthey). For him, the transformation of Monthey into *Montopia* could not be effected by 'magic' and fantasy dealing with images and representations of the city, but rather through labour and processes of restructuring. Hence, rather than rethinking urban space through fictional reality, he found it more relevant for the participants to take photos of the city and then work with these 'more real' visualizations:

> If you take pictures from, I don't know, a house or of something real you have [a] real example to do that. And if you ask in a workshop: OK this house is here, where do you think this house will be better in Monthey or anywhere? And then you can change the things there. And I think that can bring some new idea. It is maybe not possible to take a house and move over there, but maybe to change something and see in the urban something else.

It is interesting that Philippe is here creating a hierarchy of visualizations in which he considered photos of the city to be more real visualizations than the objects collected during the workshops. Even though both visualizations bore the tangible qualities of material reality, the objects that constituted the buildings and places in *Montopia* came to serve as representations of a subjective reality that, for Philippe, was less 'real' than the objective and factual visualizations of photographic images that promise unmediated access to the city. However, as Deutsche (1991) points out, visual images of the city need to be recognized as representations that imply subjectivity as well. Reality and representation mutually imply each other. This does not mean that reality does not exist, but rather that there is no one true and real city lurking behind representations and being independent of subjects. The antithesis between the forms of 'tall tales' produced in the workshops and Philippe's way of thinking is here foregrounded. Philippe's thinking can be seen in relation to a form of aphoristic thinking that moves towards transcendence and away from the immediate context of situation, seeking to subsume the situation beneath 'the rule' (Stewart 1993). The tall tale, on the other hand, is dependent on its context and the process of constant invention.

Instead of taking part in this collective project of invention, Philippe took on the role as an observer. As he explained, he saw the workshop as a useful tool for tapping into local knowledge through the transmission of information from members of the public to him as an urban planner:

> For me, it is really important to speak with everybody about everybody. And [the workshop] was a good exercise for that [. . .] It is good politics because you do not need a result on this exercise. And if you want a result, what we have to do is: if I have an idea I have to speak with people to get them on my idea and take the people with you. Here is another job: everybody gives an idea and you can hear without problem every idea, you do not have to make a decision: it is this place or this place, it is really free. But real politics is not that, it will be a lot more difficult than that. But it was interesting to talk with people that you do not know and they are voicing their opinions about places, castles, city – that was good.
>
> (personal interview, 04.09.2014)

For Philippe, the workshop provided detailed listening and enrichment. He clearly appreciated the public expressions of desires for Monthey, but at the same time he left these desires in place by regarding them as separate from any 'true' political situation in which he would have to act upon them. Accordingly, he approached the workshops as a free and open space detached from the contested realities of urban planning. The problem with this approach is that it reduces the concept of difference to the idea of a multiplicity of uniquenesses, indicating simply the acknowledgement of the existence of diverse particularities in society. The political theorists Ernesto Laclau and Chantal Mouffe (2001) call this conception 'a closed system of differences' (p. 115). Instead of seeing difference as a complex relational process in which one's subjectivity is transformed through continual identification and alienation, it is here understood as a series of distinct categories that can sometimes be held together by a broader unifying ideal. For example, Philippe was making a clear distinction between himself and the rest of the participants of the workshop: 'If you have a group like me, just like me – really pragmatic – you can't [do the same workshop] as when you are working with people that have jobs such as [Marco], people working at the theatre' (personal interview, 04.09.2014). As a result of this a priori constitution of the other participants as theatre people that deal with fiction and imagination on a daily basis, Philippe established a direct link between the participants' identity and the fantastic stories told in the workshop. This link was based on pre-existing judgements that were not only expressed by Philippe. The other participants in the workshop immediately recognized and commented upon Philippe's 'outsider' role:

Maud: He didn't play. Did you notice that? [. . .] He wanted to observe from the outside. He didn't want to get involved, you know?

Michaela: Yes, it was like we were kids and he was the father looking at the kids playing. I felt a little bit like that . . .

The authority of the 'real' urban planning over the imaginative play of the workshop was here enforced. By taking the role of 'the observer', Philippe – as representative of the 'real' urban planning – occupied the most essential relationship of authority, that between the adult and the kid. Philippe's apparent lack of engagement here served to reinforce prejudice towards unreasonable, unthinking 'authorities' that form judgements and orders in apparent detachment from the lives of the people they govern:

> It is interesting for me how some people, like [Philippe] . . . I mean they are technical and they can't see how this kind of project is not just fantasy, they are like: 'Ah artists – they are crazy! We are wasting our time.' It is frightening to see how he cannot play. But play is useful. I really think it is very . . . to see how you can connect reality and fiction, and how fiction can nourish reality. Of course it does, for me it is obvious, but it is really astonishing that for some people working with the city – I mean because he is working with that [. . .] wow . . . it is questioning . . .
>
> (Maud, 40s, theatre administrator, personal interview)

Similar prejudices were expressed in other workshop contexts. In one workshop, I noticed how a middle-aged woman repeatedly ignored the inputs from an older woman in the speed dating process. The two of them were working with a model of a public park that was created from a music stand and a red umbrella. The exercise was to imagine the place as a nightmare. The middle-aged woman immediately responded that if a basketball court were placed right next to the park, it would certainly be a nightmare as the noise would be unbearable. The older woman tried to suggest that garbage in and around the park would also be a nightmare, but the middle-aged woman kept insisting on the terrible noise from basketball courts. After the workshop had ended, the middle-aged woman approached me and explained the dynamics between her and the older woman. Apparently, the latter was a former local politician, and the middle-aged woman told me that she had had a fierce dispute with the politicians in the city regarding a basketball court right outside her house.

As these examples illustrate, the workshops produced antagonistic encounters between the participants. The productiveness of this antagonism, as advocated by Bishop, is questionable in this context. Instead of leading the participants to question and challenge fixed identities, the encounters rather served as a way of insisting on the rigidity of social roles, and thus reinforcing the antagonistic camps of 'us and them'. Equally, the workshops failed to facilitate an open dialogue, as envisioned by Kester (1999), as the pre-existing conflict of perspectives impeded a consensual understanding. Is it possible, then, to imagine a framework that would retain the power to challenge fixed or conventional meanings and identities through a dialogue that works across the 'us and them' binary?

The exhibition: testing the relation between materiality and meaning

In preparation for the exhibition, I collected all the written material from the workshops and turned them into the small descriptions of each model. For example, as an audience of the exhibition, you could read that 'Plaisir' is the biggest park in *Montopia* and is constantly filled with music and dance. Whoever attends the yearly 'play in the park audition' and makes it to the final cut of the 365 best ones is allowed to bring their instrument and play in the park whenever they like. The description of 'The House with the Elevator' – a yellow and white traffic cone that looks like a gigantic skyscraper compared to the rest of the models on the map – tells about the special smoke that comes out of its chimney, reminding the residents not to smoke cigarettes. And the story of a takeaway coffee cup that is given the name 'The Sculpture of the Last President' tells about its very important function: if you push the middle of its three eyes, you are able to speak directly to the mayor of *Montopia* and tell her your opinion about the city. The descriptions were based on my personal notes from the workshops and what the participants had said about each model, as well as the written material from the speed dating. For all the exercises, the participants had produced some form of documentation of their conversations, be it Post-it notes, fictional postcards, keywords, or the like. I had carefully marked all this material so that I knew to which model they belonged. Sabeth and Nina were in charge of the Photoshopped images. After each workshop, Sabeth marked all the places where the models had been placed on the plywood map, cycled to the places, and took pictures to be used for the Photoshopping. Meanwhile, Nina would make sure to get perfect pictures of each

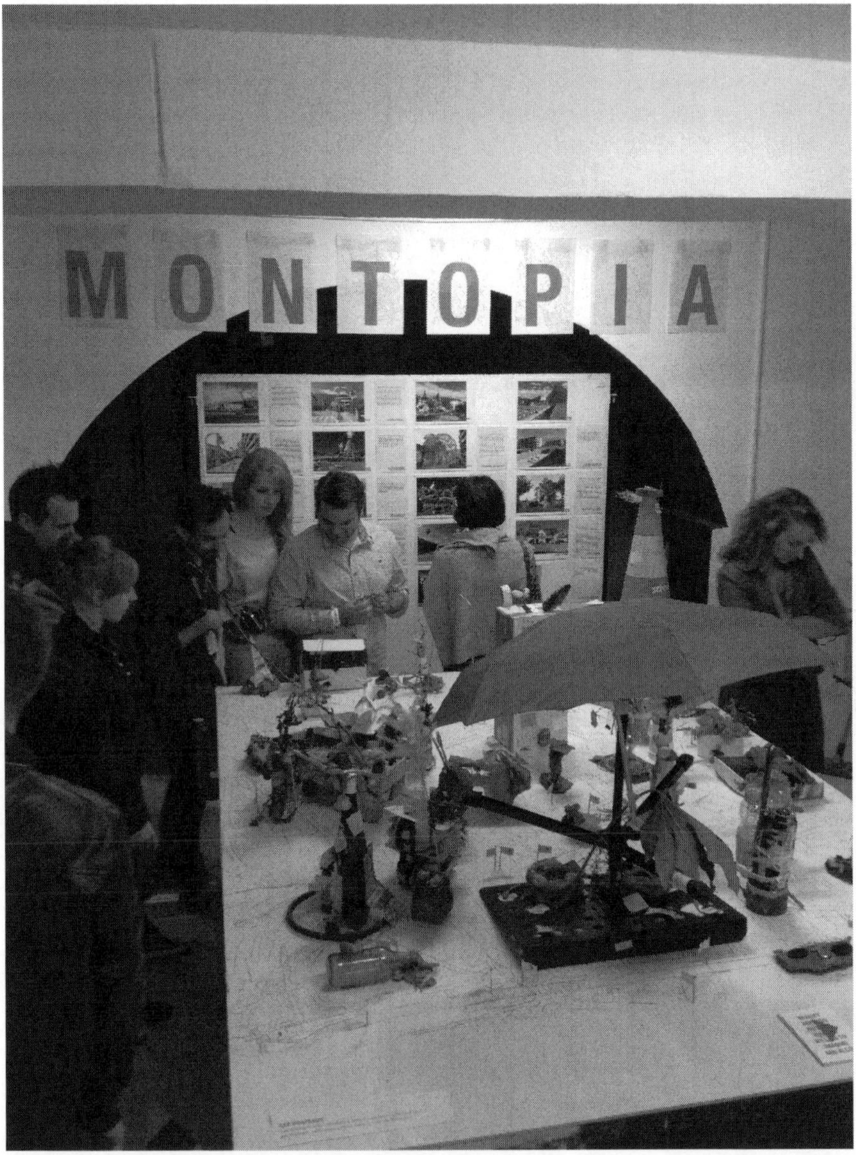

Figure 8.7 The exhibition, *Montopia*, zURBS, 2014 (image author's own).

model from all possible angles, so that it could easily be Photoshopped into the pictures taken by Sabeth. Based on my descriptions of each model, they would then produce the photo collages.

The visitors of the exhibition were enthusiastic:

> When we see the exhibition with the objects Photoshopped, then it is very cool . . . it is the result of something that is very complex. So we can think after the workshop about this town [. . .] The next step would be political action [laughs]. I joke, but it is true. It would be to, I don't know, to create participative groups to think about this city. And to make proposals – even completely irrational or impossible to do – to make projects about it [. . .] And to invite the president of the town . . . and make a lobby, a citizens' lobby about this.
>
> (Marco, 40s, theatre director, personal interview)

> I think we should write a letter to the mayor proposing to actually build 'The Sculpture of the Last President'; it's so beautiful that the kids have thought about this form of direct democracy!
>
> (Louise, 50s, exhibition visitor, exhibition talk)

> It is a way to dream of the city: you invent stories of these people, and some buildings you never looked at – you suddenly really pay attention [. . .] It nourishes your vision of the city. Definitely.
>
> (Maud, 40s, theatre administrator, personal interview)

The set-up of the exhibition mimicked that of an architectural presentation where plans are presented to the public through visual renderings, written specifications, miniature models, and technical details. Such visual projections of proposed objects or spatial alterations into the existing urban context are often used in order to demonstrate the positive, benign, or, at the very least, unobtrusive effects these alterations will have on potential physical sites. However, the *Montopia* exhibition did not aim merely to reflect or represent a glossified image in order to 'sell' the idea of *Montopia*. Rather, it was intervening in the conditions of the city's existence by turning the internal logic of the architectural presentation on its head.

As the architect Jeremy Till (1994) points out:

> The way that we conceive of and eventually make our cities, and the buildings that constitute them, is to a large extent determined by the way that we represent them [. . .] The standard method of architectural production and representation is enthralled by the classical model – stable, unified and ordered within a coherent and rational system [. . .] A typical architectural project proceeds in a steady manner from the scale of the city, through the scale of the building and finally to the scale of the architectural detail. At each of these stages, particular issues are investigated and kept within the exclusive territory of the relevant scale.
>
> (p. 239)

In this linear process, the large-scale city tends to be privileged as decisions made at an early stage determine what follows. Eventually, Till argues, the city is reduced to a series of codes that are reductive and exclusive: the scale excludes the realm of the body, the graphic excludes the social and political, and the rational method and representation exclude the imaginative, the suppressed, and the irrational. As a result, the city, as this form of master plan, is not seen as a social product of inhabitable difference, but as a system that is there to be controlled. This observation chimes with Lefebvre's and Deutsche's critique of technocratic definitions of the city as the product of experts external to its users, rather than produced by them.

The *Montopia* workshops had followed the opposite trajectory from that described by Till. Instead of starting with the large scale, the city of *Montopia* as a whole, the workshops started with the objects – the technical detail – from which *Montopia* would ultimately be built. These objects were dealt with on an intimate level, through a tactile and personal engagement. The participants were holding the objects in their hands, carefully investigating their tactile qualities (Heavy? Light? Smooth? Sticky?) and sharing personal associations relating to them (Why did they want this object in *Montopia*? How does the object relate to Monthey? What personal memories connect to the object?). At this small scale, the body is not reduced to a diagrammatic object; on the contrary, the tactile engagement with the objects enhances an awareness of one's own body in space as it fosters a sensation of how I am holding my arm or in which direction I am turning.

The next step was to turn the objects into miniature buildings and places in *Montopia*. As opposed to the rational architectural scale model where all elements are kept within the exclusive territory of the relevant scale (e.g. people, trees, and cars are made in the same scale as the buildings), the *Montopian* models were created by transforming the context and not the structure of the objects: the bottle becomes a fountain, the traffic cone becomes a hotel, and the mushroom becomes a climbing gym. Through this process, the objects became animated, and thus initiated another world, a world that cannot exist without the eye performing certain operations, manipulating and attending to the physical world in ways that test the relation between materiality and meaning (Stewart 1993). The miniature model of *Montopia*, then, resisted the formal ordering system of the architectural model, and thus also the distanced position of the viewer. Whereas the architectural model offers a transcendent view of an enclosed rational and technical whole, the *Montopia* model made explicit its representative quality by testing the relation between materiality and meaning, and thus offered itself as a stage on which the viewers could project a series of actions, suggesting use, implementation, and contextualization.

The visual renderings in which the *Montopian* models were Photoshopped into present-day Monthey aimed to make a plausible reality out of the possible worlds offered by the model. In these images, the objects were magnified to disconcerting effects; the traffic cone looks tall as a mountain, a flower covers the whole roof of a building block, a glass bottle is the size of a house, a feather takes the shape of a huge sail, a pine cone is twice as big as the excavator in front of it, and so on. The idea of the gigantic is here evoked. Stewart (1993) points to how our relation to the gigantic is articulated in our relation to landscape as we move through it and are enveloped by it. Our position here is the antithesis of our position in relation to the miniature: whereas we know the miniature as a spatial and closed whole, and we thus observe it from a distant position, the gigantic envelops and surrounds us. The awareness of the body in space is here heightened as the body remains the constant measure of our articulation of the scale of the objects. The miniature *Montopian* models on the plywood map could be held within our hand, but the hand is not in proportion with the *Montopian* world; instead, the body becomes some sort of background landscape. This position of the body is juxtaposed in the Photoshopped images where we are enclosed in the shadow of the gigantic objects. The gigantic here presents a physical world of disorder and disproportion, in contrast to the mental world of proportion, control, and balance as represented by the architectural model. Accordingly, the narratives of the miniature and the gigantic mutually define and delimit each other.

Parallels can here be drawn to the surrealist collages as mentioned in Chapter 3. In both instances, juxtaposition is used to disturb and defamiliarize any notion of wholeness and unity. The *Montopia* models portrayed in a painstakingly lifelike fashion buildings and places that do not belong to our shared perception of external reality, so as to convey the experience of an alternative world that is at the same time dreamlike and absolutely real. The juxtaposition between the two states of dream and reality created what Clifford (1988, 133, cited in Pinder 2005, 120) calls 'the order of an unfinished collage rather than that of a unified organism'. In a surrealist vein, then, the Photoshopped images defamiliarized the everyday, and drew attention to the urban landscape as dynamic arrangements and organizations of disparate parts and objects. This approach reflects the surrealist take on utopia as experimental rather than conceptual. Experimental in the sense that utopia involves processes of change, and finding new places in the present or new relationships with place in a playful manner, rather than constructing blueprints for the future (Fenton 2006).

The written descriptions accompanying each Photoshopped image were continuations of the tall tale-like stories told by the participants during the workshops. By placing these narratives in an exhibition setting, which appropriated the domain of authority of 'architectural pin-ups', the narratives undermined the discourse of 'reality' by usurping it, and thus aimed to transcend the real/unreal dichotomy. The language of the written descriptions shuttled back and forth between the language of imagination and the language of experience, articulating utopian tendencies that we cultivate from childhood when inventing magic kingdoms and narratives to express our wishes and desires. The fact that some of the buildings and descriptions for *Montopia* were actually made by children, but that it was not clear which ones, made this link more explicit.

The uncertainty about which groups of participants made which models challenged the notion of fixed and stable identities that was manifested in the workshop process. Visitors to the exhibition frequently asked questions about what models and narratives belonged to which groups of participants, surprised to learn that the interactive 'direct democracy' sculpture was made by the kids while the huge beach-cinema complex was made by the architects, that 'the crazy city' which functioned as a party place was made by the parent group while the 'all are welcome' sign for *Montopia* was made by the teenagers. Thus, the exhibition epitomized what John Dryzek and Simon Niemeyer (2008) define as a 'discursive representation' that represents various discourses rather than particular individuals or groups. A discourse can here be understood as 'a set of categories and concepts embodying specific assumptions, judgements, contentions, dispositions and capabilities' (p. 481). Discourses should not be confused with groups defined by ascriptive characteristics such as race, class, or gender. Individuals and groups can inhabit multiple, and sometimes even countervailing, discourses, and hence what is represented is not the participants as whole people, but some of their interests, identities, and values. By fostering this form of 'discursive representation', the exhibition had the potential to destabilize (group) identities, thus initiating a form of critical reflexive consciousness regarding the limits or 'finitude' of community (Bishop 2012; Kwon 2004; Nancy 1991; Young 1986).

Conclusions

As this chapter illustrates, the contradiction between the arts and the social was hard to negotiate in the participatory workshop process of *Montopia.* Whereas the workshops had the capacity to operate on a level removed from the everyday, and occasionally did enable some participants to explore and express desires for a better city, they did not provide a 'free' aesthetic field completely removed from domestic and domesticated modes of sociability. Everyday prejudices relating to the participants' predefined social situations and identities in the city led to antagonistic encounters that reinforced the binary camps of 'us and them'.

The exhibition was more successful in negotiating the contradiction between the arts and the social. Instead of trying to ensure a direct link between the workshops and the 'reality' they were seen to represent, the *Montopia* exhibition undermined the discourse of reality as something the workshop referred to and presented it rather as something it produced. The exhibition recognized the role of the aesthetic as a distinct sphere, but at the same time attempted to work towards an egalitarian form of social relations by opposing totalizing narratives that claim to represent 'real life'. The contradiction between art and the social was here a fruitful starting point for the exhibition, rather than some kind of negation of it. In a surrealist vein, we juxtaposed the fantastic *Montopia* models with everyday urban space in Monthey so as to challenge the perceived boundary between the fantastic and real, the arts and the social. The idea of any whole or unified representation – whether of *Montopia* or Monthey, or of the participants' roles in the workshop or their roles in the everyday – was here destabilized and challenged. It became clear that, in this regard, the exhibition had succeeded where the workshops had not.

However, we did not want the participatory workshop process to be seen merely as a preparatory step for the exhibition. The participants should not feel that we had 'used' them solely as inspiration for our own work with the exhibition. We feared that our artistic adaptions of the participants' contributions would compromise their sense of ownership of the project. At the same time, we felt that these adaptations were necessary in order to be able to communicate the project to a larger audience. How could we find a balance between ethical responsibility and artistic quality in this regard? How could we ensure autonomy in what we do and at the same time be ready to compromise and carefully negotiate the artistic content? With our next project, *St. Clement's Utopolis*, we tried to address these questions.

Bibliography

Bishop, C. 2012. "Participation and Spectacle: Where Are We Now?" In *Living as Form: Socially Engaged Art from 1991–2011*, edited by N. Thompson, 34–45. New York: Creative Time.

Clifford, J. 1988. *The Predicament of Culture: Twentieth-Century Ethnography, Literature and Art*. Cambridge, MA: Harvard University Press.

Deutsche, R. 1991. "Boys Town." *Environment and Planning D: Society and Space* 9: 5–30.

Dryzek, J. S. and Niemeyer, S. 2008. "Discursive Representation." *American Political Science Review* 102 (4): 481–493.

Fenton, J. 2006. "Contemporary Surrealist Geographies of Utopia and Dystopia." In *Imagining the City, Volume 1*, edited by C. Emden, C. Keen, and D. Midgley, 273–294. Bern: Peter Lang.

Kester, G. 1999. "Dialogical Aesthetics: A Critical Framework for Littoral Art." *Variant 9 Supplement* 2 (9): 82–123.

Kwon, M. 2004. *One Place after Another: Site-Specific Art and Locational Identity*. Cambridge, MA: MIT Press.

Laclau, E. and Mouffe, C. 2001. *Hegemony and Socialist Strategy*, 2nd edition. London: Verso.

Nancy, J.-L. 1991. *The Inoperative Community*, edited by P. Connor. Minneapolis, MN: University of Minnesota Press.

Pinder, D. 2002. "In Defence of Utopian Urbanism: Imagining Cities after the 'End of Utopia.'" *Geografiska Annaler: Series B, Human Geography* 84 (3–4): 229–241.

———. 2005. *Visions of the City*. Edinburgh: Edinburgh University Press.

Stewart, S. 1993. *On Longing*. Durham, NC: Duke University Press.

Till, J. 1994. "The Urban Miniature." In *The Urban Scene and the History of the Future*, Proceedings of the ACSA European Conference, London, pp. 239–241.

Young, I. M. 1986. "The Ideal Community and the Politics of Difference." *Social Theory and Practice* 12 (1): 1–26.

9 *St. Clement's Utopolis*

A crisis of identity – art or social work?

From arts festivals to an urban development project

In June 2015, we – the zURBS team – made an attempt at repeating the workshop and exhibition process of *Montopia*, now renamed as *St. Clement's Utopolis*, in London. This endeavour was based on a call that was distributed to the mailing list at the university department where I was working in February 2015. The call was for local collaborators for a so-called 'public art project' that was to be part of the redevelopment of the St. Clement's Hospital site on Mile End Road in London. St. Clement's used to be a well-known mental hospital in Mile End, featuring in famous literary novels such as Winfried G. Sebald's book *Austerlitz*. The buildings were originally built in 1848–1849 as workhouses, but became an infirmary for the City of London Union in 1874 and a psychiatric unit in 1936. Although undergoing various institutional changes, the hospital remained a psychiatric unit until it was closed in 2005. In 2011, the mayor of London sold it to the developer Linden Homes for them to turn it into a residential complex of 252 dwellings. Following 10 years of campaigning by local residents and organizations for an East London Community Land Trust (ELCLT), which attracted backing from former London mayors Ken Livingstone and Boris Johnson, the site was in 2012 declared the UK's first ever urban community land trust for affordable homes. After further negotiations, 71 of the new dwellings were set to be 'affordable' in different ways. The Peabody housing association is managing 58 of these dwellings, making them available for low-cost social rent, while ELCLT will manage the remaining 13 for low-cost ownership.

It was in the context of this regeneration scheme that the public art project was commissioned. As the consultant commissioner articulated it, '[Linden Homes] are required to provide some public art as part of the planning requirements. They have appointed me to sort it out for them' (email to university mailing list, 05.02.2015). Accordingly, the consultant had commissioned a local blacksmith to make permanent public artworks for the site: iron gates and wayfinders that would help residents of the new homes orient themselves in the area. However, as the consultant argued in her email, 'an important part of any public art project is that it should be public'; thus, in addition to collaborating with a local storytelling company, she was looking for other local resources to work on the community aspects of the project.

We had mixed feelings about being part of a so-called 'public art project'. zURBS had previously been involved in the public art festival Art and the City, which was the largest festival for public art ever organized in Zurich. The festival was organized by the city of Zurich and involved displaying 43 outdoor artworks by world-famous artists such as Ai Weiwei in public space. We had considered taking part in this festival as an opportunity to critique – from the 'inside' – the notion of 'public art' as promoted by the festival. Using public space as a form of outdoor gallery for imported artworks did not chime with the aim of the festival to use art as a way to 'seek out dialogue', 'ask questions', and 'create an identity' as stated in their press release. Rather than creating a debate, the festival seemed to try to decorate urban space as part of a wider city-branding strategy (e.g. see Evans 2003; García 2004; Quinn 2010). We had similar concerns about the 'St. Clement's public art project'. The 'sorting out' of the public art requirement for the developers made it sound as if we would again be cast as a 'participatory alibi' to meet the demands of governments and funding organizations for urban development schemes to be socially and locally inclusive.

Yet, in spite of the somewhat problematic framework of the project, we were not blind to the possibility that our scepticism about and resistance towards this type of public art project were better enabled by participation in the system rather than an escape from it, as argued by Joseph (2002). Furthermore, it was appealing for the first time to be working outside the artistic framework of the arts festival and cultural institution and to get directly involved in an urban redevelopment project. Hence, we replied to the call, proposing to do the project *St. Clement's Utopolis* – a version of *Montopia* adapted to the specific context of the St. Clement's site. Along the lines of *Montopia*, the idea was to involve local community groups in workshops in which they would build an imaginative model of the St. Clement's site in order to make it the area of which they had always dreamed. The response from the consultant commissioner was enthusiastic, and zURBS was promptly included in the project. After presenting our project to all partners (the blacksmith, the storytelling company, and representatives from ELCLT and Linden Homes), the framework for the collaboration was established: zURBS' workshops would inform the making of the final public artwork by the blacksmith-artist. The blacksmith would thus take part in all the workshops, and we would include some exercises where the participants would do 'wirework' in the speed dating process. The wirework was a method used by the blacksmith where she would bend wires to make two- or three-dimensional motifs that she later could translate into larger ironworks. The local storytelling company had a particular interest in the stories that would be told during the workshops and suggested that they would film the workshops. The film would then document the process, as well as some of the local stories present in the area. The participants would be 'recruited' through the networks of ELCLT and the storytelling company. We were excited to find a similar support structure in this project to that provided by the festival organization in Monthey, in terms of the local networks and participant recruitment. Also, the possibility of someone making a movie of our work would be valuable documentation and PR material

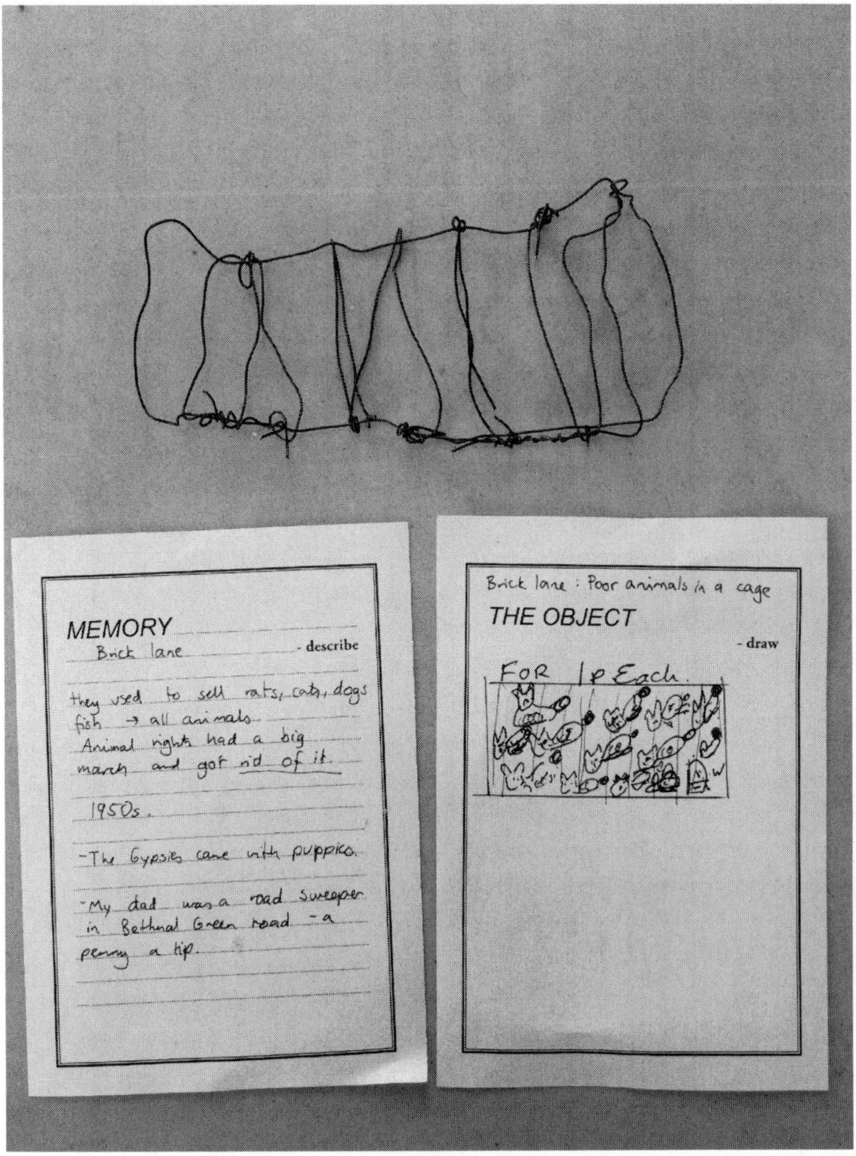

Figure 9.1 Wirework by participant, *St. Clement's Utopolis*, zURBS, 2015 (image author's own).

to have in the future. We were also looking forward to working with the blacksmith and being part of creating a participatory and permanent artwork for the site.

However, as much as everything seemed to be set for a great collaboration, *St. Clement's Utopolis* posed a series of challenges for zURBS. The project led us into an identity crisis that threatened to dissolve our whole practice and collective. By entering what Bishop (2006) calls 'the slew of community-based practices', the fine line between social work and artistic practice became so blurred that we lost track of the conceptual density and artistic significance of our project in favour of keeping a good relationship with our collaborators. Through *invisible Zürichs, stadtARCHIV*, and *Montopia*, zURBS had carefully elaborated a set of methods, frames of reference, stylistic characteristics, and attitudes towards social and aesthetic issues. This distinctive voice and outlook was put to the test in each of our projects, over and over, because it had to be related to the specific context of each project. In some cases, we chose to complicate our position as authors, or share it with others, but that choice was made by us as authors with a clear motivation and autonomy. In *St. Clement's Utopolis*, this clear motivation and autonomy were deeply challenged. More than ever, we felt caught between a generalized set of ethical precepts and the artistic specificity of our practice.

zURBS as a public art project?

The partners of the St. Clement's project had clear ideas of which local institutions to involve in the project. They found it important to engage so-called marginalized groups such as the local Bangladeshi community and residents from a home providing accommodation facilities and welfare services for active and retired seafarers, ex-army and ex-service personnel, and homeless adults (LAFWS). As a representative for zURBS, I made it clear that the wide spectrum of participants that we envisioned included not only marginalized groups, but all kinds of resident groups. We did not want to stigmatize certain groups as being in particular need for our workshops, but rather to build up a relationship with our participants based on a mutual interest in exploring the local neighbourhood. However, in the 'matchmaking mediation' (Kwon 2004) undertaken by our partners, it became clear that their motivation was based on a presumption of zURBS' interest in working with marginalized groups and the anticipation of a particular kind of collaborative project. The problem with these presumptions was that they reduced, and maybe even stereotyped, the identity of zURBS and the community groups. The 'community' here became a synonym for the social groups of the marginal or underprivileged classes. Four workshops were accordingly arranged with residents of a sheltered housing for the elderly, residents from LAFWS, 'learners' from a centre providing drop-in activities for immigrant women, and finally schoolchildren from the local elementary school.

Kwon (2004) points to how certain types of community groups are often favoured for artistic partnerships because of the easy correspondence between their identity and particular social issues. Involving, for example, the homeless residents of LAFWS would fuel the branding of the housing project as socially inclusive. The 'easy' correspondence between the identity of LAFWS as marginalized and the social concern with 'inclusion' is here put to the fore. This is problematic as it represents the primary division in society as one between an included majority and an excluded minority. The goal of the public art project is here simply to provide a transition across the boundary from excluded to included, to allow people to engage with the St. Clement's site – even if they did not have a direct relation to the site as such. As one of the residents of LAFWS replied when I asked if he would take part in the workshop, 'What's in it for me? It's not like the housing that is going to be built at St. Clement's will be available to me'. Kwon (2004) sets up the following equation for this form of community-based art: 'artist + community + social issue = new critical/public art' (p. 146). Here, the identity of a community group serves as the thematic content of the artwork, representing this or that social issue, as well as the community itself, in an isolated and reified way.

As a public art project, zURBS' task was accordingly to perform some form of social work in terms of facilitating a transaction between these vulnerable community groups and the St. Clement's site. This transaction had the perspective of problem-solving in terms of a reversal of hierarchies (e.g. giving voice to the underprivileged) and simultaneously providing a certain compensation for them (Riet et al. 2012). This became clear in an email conversation where the consultant commissioner, the ELCLT representative, and I discussed a PR text for disseminating information about the workshop to potential participants. Following zURBS' aim of having the participants question, rather than blindly accept or come to terms with, the developments happening in their local area, I proposed a text that emphasized the aim of letting the participants articulate their desires and wishes for their neighbourhood through all possible imaginary projections for the St. Clement's site. However, the representative for ELCLT kindly suggested we should change the conceptual framework of the workshops:

> I wonder whether we should introduce the workshops more along these lines: 'In the context of the redevelopment of St Clement's hospital site, we are inviting you to participate in helping us to design some public art for the development . . .'? Many of the people we will be approaching have already been involved in community design workshops, which helped the architects to refine how the plans for the site were developed. We don't want to give the impression that we are repeating that exercise.
>
> (personal email, 09.03.2015)

This reconceptualization of the workshops put zURBS in an awkward position. We had to accept that we could not do an exercise that would be perceived as similar to the community design workshop when the construction was already well underway. This would completely undermine the initial efforts at public engagement. At the same time, it was unclear to us how these workshops had taken place – whether we were talking about completed proposals being represented for 'feedback' in which the only agency left to the public was the simple act of registering its approval or disapproval. Furthermore, we risked replicating this latter approach by offering a site for simply 'incorporating' community groups into the development rather than providing them with an opportunity to question or challenge it. How could we find a middle way that did not undermine the previous attempts at public engagement, but at the same time opened up a space for critique?

The framework of letting the participants design their own proposals for the public artwork for the St. Clement's site seemed like a potential for critique in this regard. We encouraged the participants to make public artworks that could be seen as monuments to their personal relationships with the area. We emphasized that even though the physical infrastructure of places changes, people's memories of and associations with these places remain. This perspective was an attempt to create a shared sense of ownership among the participants, and to point to possibilities for appropriating space in spite of major structural transformations. Nevertheless, our ideological aims were somehow contradicted by the fact that zURBS was collaborating with the developers of St. Clement's, and thus could easily be seen to represent the interests of the developers in terms of 'selling' the project rather than critically scrutinizing it. In this regard, the workshops could be critiqued for offering quite passive and static roles for participants, defined by the need for inclusion rather than the achievement of specific purposes or goals.

Reaffirming preconceived identities

In developing the workshops, we were faced with a number of demands, relating to ethical and practical considerations, from the institutional administrators responsible for the community groups. When trying to navigate in this complex web of demands, motivations, expectations, and projections, we increasingly felt that we had to prescribe the nature of the groups' participation in advance of the workshops. The older people would be more than happy to share their memories of London, we were told by the administrator of the home. Hence, we should focus on letting individuals tell their personal stories in the workshops. The speed dating concept and collecting objects in the city should be skipped as the older participants wouldn't 'have the ability or inclination to do that'. The 'learners' of the women's centre were very insecure about the English language and also in terms of interacting with strangers, the administrator told us. Hence, we should start the workshop with an informal 'get to know each other' chat (preferably revolving around the topic of children and family) in order to create a familiar and safe frame around the workshop activities. The storytelling company made it clear that the residents of the LAFWS would probably be the most challenging group to work with. This was confirmed by the administrator of the home, who advised us to accompany the participants on their hunt for objects as they most likely would be too disengaged to carry out that or any other task on their own.

We did our best to tailor each workshop to these specifications and considerations. However, in determining the nature of the collaborative relationship between the participants and ourselves, we overlooked the discursive constructions of our participants' identities. We fell in the trap that Kwon (2004) warns against, and considered the objectives and identities of the different participant groups as determined before any encounter with outside individuals or groups such as zURBS. Kwon points to how this form of 'essentializing process' enforces the idea of community as a unified and existing social relation defined through the isolation of a single point of commonality and a presumption of unified subjects. The identity that is created through the art project, then, is viewed as a self-affirming, self-validating 'expression' of a unified community neglecting the representative frameworks that direct such productions of identity. Accordingly, rather than focusing on how the workshops could help engender different types of identities, we focused on the self-expression of the preconceived identities of the participants: the elderly were

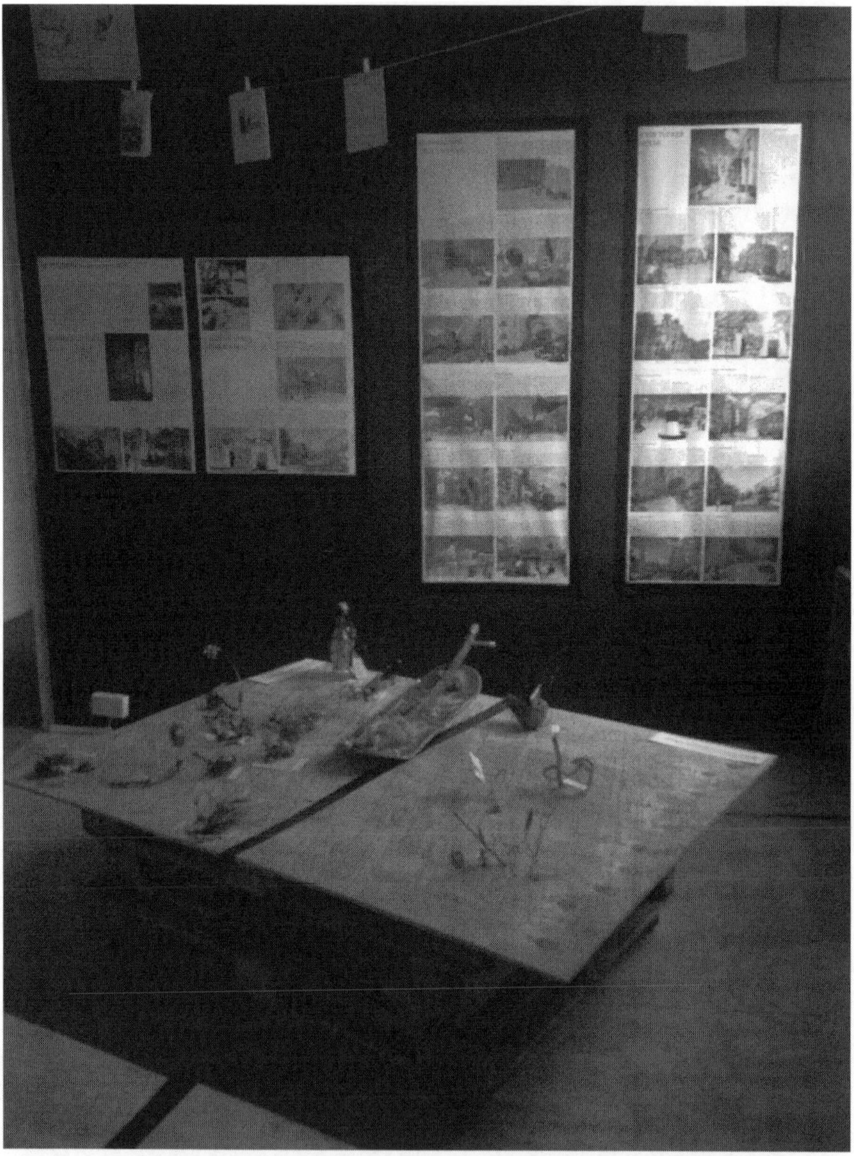

Figure 9.2 The exhibition, *St. Clement's Utopolis*, zURBS, 2015 (image author's own).

encouraged to share their memories and experiences from the past, recounting stories about how they experienced East London during the war or about how they arrived in London as young adults in search for better lives; the schoolchildren were asked about the nearby parks and the playgrounds and where they liked to hang out; the women's group was invited to discuss their children and the cultural traditions of their home countries; and the residents of the LAFWS were accompanied out into the streets, where we tried to facilitate discussions revolving around their experiences of the struggle to live and find work in the area.

There is, of course, nothing wrong with the stories stemming from these self-expressions. They are indeed important for giving voice to the underprivileged and clarifying the personal and moral meaning of social issues and events. The problem was the framework through which these stories were told and subsequently represented. For example, the blacksmith used the stories as inspiration for motifs in her ironworks for the St. Clement's site. These motifs could be critiqued for being stereotyped and simplified: a motif of various religious symbols represented the different cultural traditions talked about within the woman's group, an old-fashioned bell represented the stories told by the elderly, an anchor represented the residents of the LAFWS, and a bicycle represented the schoolchildren. The motifs were exhibited in a public exhibition together with the models, photo collages, and descriptions from the workshops. As opposed to the *Montopia* exhibition, where the authorship of the models was obscured by the fact that all participants had placed their models on the same plywood map, the models in *St. Clement's Utopolis* were placed on separate plywood maps according to the participant groups that had made them. Similarly, the stories and photo collages were divided according to the authorship of the participant group.

The choice to make the authorship of the models explicit related to what we perceived as the wish of the other project partners to make visible a certain form of authenticity in which each model, as a representation of one particular community, was ostensibly produced by that same community. Whereas the *Montopia* exhibition resisted any coherent representation of collective identities and unsettled the status of the 'other', the *St. Clement's Utopolis* exhibition sought to display a variety of social and cultural differences, as represented by the different participant groups, at the expense of a critical examination and questioning of these differences. Hence, the question is: When the process of this constant negotiation between different interests, agendas, and expectations between institutions, participants, and artists is over, who has actually spoken and what was it possible to say?

Enabling critique through transformative experiences

> I want to be more explorative, more urban-researchy . . . (hard to find the proper word – let's say: more zURBS-style!). However, I always felt that this wasn't REALLY working, because something was missing. Some playful moment, maybe, some more performative aspect that puts [the participants] into the necessary state of mind/being. (btw: I DON'T think that performance

is elitist at all. The way I see performance supplementing zURBS-projects, it's rather the opposite, it's the bridge that makes something more approachable, more interesting, a bit less serious, a bit less academic, and certainly less school-style.) That's what I meant. When working with the lay-people, we need to work MUCH harder, to make what we're making well. Because we cannot expect them to enjoy it because they UNDERSTAND what we're aiming for (as it seems to work in academic contexts). They're not even supposed to understand what we're aiming for (which is: making them look, re-look, re-think, imagine and finally emancipate themselves from what they thought, from what they're supposed to see/understand/do . . .). It should be something that just HAPPENS – happens THROUGH our projects. And for that, we're not good enough yet, in what we're doing. We're good, and it DID happen every once in a while, but in order for me to be confident about our projects, it has to work better in that sense. And that is sort of the condition for me to actually stand there and present it, convinced (and proud) of it. Convinced that they will have experiences which are worth the participation.

(personal email from Sabeth, 31.08.2015)

As mentioned, *St. Clement's Utopolis* made us – the zURBS team – seriously doubt our approach and artistic integrity. In negotiating with the different project partners, we had not only compromised the identities of our workshop participants and predetermined the nature of their participation, but we had also failed to give our participants 'experiences which were worth the participation'. The notion of 'experience' needs some further scrutinizing as it cuts to the core of the questions raised in this chapter in terms of finding the balance between ethical responsibility and artistic quality.

Thomas Lemke (2011) argues that Foucault, in his later works, sought to base the idea of critique on the notion of experience. Foucault conceives of experience as a multilayered concept that articulates forms of knowledge, mechanisms of power, and relations to the self. The link between experience and critical activity is here put to the fore:

> The experience through which we grasp the intelligibility of certain mechanisms (for example, imprisonment, punishment, and so on) and the way in which we are enabled to detach ourselves from them by perceiving them differently will be, at best, one and the same thing.
>
> (Foucault 2000, cited in Lemke 2011, 29)

Experience is not understood as a negative practice that necessitates the determination of rational standards of evaluation and the application of these standards to social reality. Rather, Foucault suggests a more positive activity that problematizes the way we think about and judge certain objects in order to distance ourselves from their naturalness or self-evidence, and to work towards new experiences. Critique is seen as an instrument or a means to achieve something that does not yet exist: the promise of a better future (Lemke 2011, 30). This preference for a form of critique that focuses on experience resonates with Sabeth's email and the general aim we had for the workshops of being valued not as a demonstration of certain groups or perspectives, but for the affective experience that the workshop made possible for the participants. This form of experience is always a fiction since it is something one fabricates oneself, and thus doesn't exist before, but will exist afterwards. Hence, it has the potential to move beyond the limits of the present. The artistic framework of our workshops played a crucial part in enabling these forms of experiences. The simple act of blindfolding and listening to the sound clip in the elevator inviting the residents of Monthey to enter *Montopia* fuelled the participants' imaginations. Occasionally, these acts provided experiences that transformed the participants' perceptions of the city, and inspired them to more local activities that sketched out the potentials of these transformed perceptions, such as proposing to the mayor that the *Montopian* 'Sculpture of the Last President' should actually be made.

In the *St. Clement's Utopolis* workshops, these kinds of transformative experiences were heavily compromised. Out of considerations for institutional concerns about the perceived 'vulnerability' of the workshop participants, we focused on providing a familiar and safe setting that corresponded to a set of 'considerations' that was articulated by the institutional administrators. These considerations led us to neglect the aesthetic and performative framework of our workshop by skipping blindfolds, soundscapes, and role play. Instead, we focused on how we could best communicate the project in terms of involvement in the St. Clement's site and the development of the public artwork, and at the same time how we could demonstrate an awareness of the problems and challenges potentially faced by the participants in this regard. Our ethical concerns were here starting from moral concerns or obligations posed by our institutional collaborators, resulting in us framing the workshops in ways that would comply or conform to pre-established norms and regulations. As a result, the transformative experience of the participants became restricted and weakened. Instead of allowing space for new subjectivities to emerge, the ethical considerations articulated by the institutional administrators defined the limits of the participants' identities and possible actions in the workshops. Hence, the workshops did not have 'the function of wrenching the subject from itself' (Foucault 2000, cited in Lemke 2011, 36).

Transgressing the limits of who I am cannot, according to Foucault, start with the formation of an identity. Rather, an identity can only be the result of elaborating on the question of what or who one is, as it is posed through a process of transformative experience that provides the new terms in which one formulates it. In other words, experience calls for the formation of new subjectivities that may be constituted in disobedience to the principles by which one is formed. From this perspective, the ethical responsibility of zURBS was to enable this form of disobedience through, for example, the sensuous terrain of affect as mentioned regarding Sigman's work in Chapter 3. According to Thompson (2009), the attention to 'affect', defined as bodily responses, sensations, and aesthetic pleasure, should be the basis of an ethical focus for arts processes, as it may expand the participants' capacity to act, engage, and connect. Affect is here seen to intensify experience in a way that can both protect and draw the participants into the world around them, without insisting on the terms of that engagement.

Conclusions

As I have illustrated in this chapter, the potential of articulating alternative urban futures was lost in the *St. Clement's Utopolis* workshops. In order to comply with pre-established norms and regulations from the various institutional collaborators, we had to reduce the aesthetic and performative framework of our workshops to an absolute minimum. As a result, we limited the participants' possible actions in the workshops, and failed to provide them with any transformative experience that would allow for new subjectivities and discoveries to be made. In order, then, to enable critique through transformative experiences, it is necessary to rethink the ethical responsibility of socially engaged art. Instead of focusing on ethics as a regulative power that tries to limit or control frames of participation and representation based on a moral responsibility relating to a series of predefined requirements and obligations, socially engaged art calls for an understanding of ethics that destabilizes these frames. This can be seen as a middle way between Bishop's and Kester's dispute regarding the contradictory nexus of collaborative artworks, where the transformative potential of art directly confronts its institutional character. Rather than abandoning ethics altogether (as advocated by Bishop) or 'sacrificing' aesthetics in favour of ethics (as advocated by Kester), this approach encourages careful negotiation of, and reflection on, the subtle power relations that exist between artists and their collaborators (Charnley 2011). Ethics is here not a process of making a 'free' space that increases the participants' ability to act unencumbered by the restrictions imposed by external forces. It is instead a process of recognizing the direct and felt responsibility of the experience of engaging and being with others. Accordingly, socially engaged art may facilitate critical activity as a mode of self-formation that suspends as far as possible the normative system that one refers to in order to test and evaluate it (Lemke 2011). Understood as a destabilizing frame, ethics may help artists and researchers to question our methods and approaches, rather than function as a set of moral codes to be applied to practice on the base of institutional demands and constraints. In the Conclusion of this book, I will further elaborate on how this destabilizing frame may provide broader reconfigurations of socially engaged art practice by working with, rather than against, the inherent contradiction of this form of collaborative art.

Bibliography

Bishop, C. 2006. "The Social Turn: Collaboration and Its Discontent." *Artforum* 44 (6): 178–183.

Charnley, K. 2011. "Dissensus and the Politics of Collaborative Practice." *Art & the Public Sphere* 1 (1): 37–53.

Evans, G. 2003. "Hard Branding the Cultural City: From Prado to Prada." *International Journal for Urban and Regional Research* 27 (2): 417–440.

Foucault, M. 2000. "Interview with Michel Foucault." In *Essential Works of Michel Foucault, Vol 3: Power*, edited by M. Foucault and J.D. Faubion, 239–297. New York: The New Press.

García, B. 2004. "Cultural Policy and Urban Regeneration in Western European Cities: Lessons from Experience, Prospects for the Future." *Local Economy* 19 (4): 312–326.

Joseph, M. 2002. *Against the Romance of Community*. Minneapolis, MN: University of Minnesota Press.

Kwon, M. 2004. *One Place after Another: Site-Specific Art and Locational Identity*. Cambridge, MA: MIT Press.

Lemke, T. 2011. "Critique and Experience in Foucault." *Theory, Culture & Society* 28 (4): 26–48.

Quinn, B. 2010. "Arts Festivals, Urban Tourism and Cultural Policy." *Journal of Policy Research in Tourism, Leisure and Events* 2 (3): 264–279.

Riet, S., van Eeghem, E., Verschelden, G., and Dekeyrel, C. 2012. *Reading Urban Cracks: Practices of Artitsts and Community Workers*. Ghent: MER. Paperkunsthalle.

Thompson, J. 2009. *Performance Affects*. London: Palgrave Macmillan.

Conclusion

Creating a new course for socially engaged art within the neoliberal city

In autumn 2015, Nina, Sabeth, and I spent one week together in the Swiss mountains to discuss zURBS' future. The perceived 'failure' of the St. Clement's workshops had triggered a lot of insecurities and disillusions regarding our practice. Our doubts were in many ways quite paradoxical. On the one hand, we were convinced like never before of the importance of criticizing, questioning, and challenging present technocratic approaches to cities and urban space. As I have illustrated in this book, our practice with zURBS had made visible how technocratic and neoliberal visions to a large degree dominate how we live in, imagine, and think about our cities in relation to processes of meaning-making, participation, and representation. Our behaviour and practices, as citizens, in urban space are supported by material forms and designs that play an active role in directing our actions within it. The material environment is often portrayed as an organization of fixed and neutral physical objects that 'just is', and thus is not up for negotiation in terms of, for example, questioning the social and political interests, speculations, and calculations that produce it. Instead, our material surroundings are designed to gloss over growing structural problems relating to, for example, inequality and accessibility, and become a powerful force in rendering urban inhabitants politically passive. zURBS' aim of challenging what is presented to us as given, by enabling urban inhabitants to identify and positively and creatively mediate power imbalances attached to social, material, and imaginary aspects of urban space, appears crucial within this context.

However, acknowledging the importance of what we were trying to do made us question whether the imaginative interchange and open-ended, exploratory processes that we fostered through our projects were enough to generate change. To repeat the Harvey (2000) quote from the beginning of this book, the future needs to be constructed out of the tangible materials of the present, and not solely in some 'fantastic utopian mold' (p. 191). Wouldn't we better approach change by using our artistic skills and efforts in campaigns, demonstrations, and actions against policies and development schemes that are based on private property and exchange value? Wouldn't it be more effective to set up a community garden or local neighbourhood activity group if we wanted to enhance participation and social relations among urban inhabitants? What was the point of socially engaged artistic interventions if they were not accompanied by structural change in the world at large? Should we dissolve zURBS and either use our energy in direct social action and campaigns or unfold our creative urge in art projects that were not necessarily socially committed?

These questions point to the inherent contradiction between aesthetics and ethics in socially engaged art, in terms of attempting to cross the boundary between art and the social, and at the same time aiming to stand apart from the social as a distinct critical space. As a way of concluding, in what follows, I discuss this contradiction in the light of the experiences I have outlined in this book. Whereas I point to the challenges this contradiction poses, I will also illustrate how it might provide a direction towards broader reconfigurations of socially engaged art and generate a desire for change. Based on these reconfigurations, I will offer a model for how socially engaged art can productively mediate between urban imaginaries and material environments, and this way combine utopian aspirations and concrete action in order to challenge conceptions on the production of urban space within the neoliberal city.

The pervasiveness of contradiction

In this book, I have combined geographical research and artistic practice in order to offer a close empirical investigation of the entanglement between socially engaged art, its participants, and its social and material site. As I have illustrated, this entanglement is ridden with contradictions. This is nothing new; contradiction is inherent in human existence and embedded in its history. As Harvey (2000) argues, any mode of production is a contradictory and dynamic unity of different elements. However, many discussions of socially engaged art fail to carefully examine the various contradictions produced within and through it. As discussed in Part I, this is because these discussions often focus solely on the subject-centred relations that are produced through this form of artistic practice, and do not carefully examine the interdependence between these social relations and their material context. This book provides a more comprehensive analysis by combining critical ideas about the social construction of art with perspectives taken from critical urban studies such as the power and politics of producing (urban) space. The value of this analysis is that it problematizes the idea of socially engaged art as an inherently liberatory space and reveals the often-suppressed contradictions within artistic as well as urban processes.

Accordingly, in Part II, I foregrounded how the city should be seen not as an assemblage of neutral objects produced by experts, but as an environment formed through various social practices that involve all its users. The artistic practices of Wentworth, Sumégne, and Sigman here provided concrete examples of how the material environment and its associated meanings are played out as parts of social strategies. By drawing attention to how we use, understand, and give meaning to urban space through our critical practices within it, Wentworth, Sumégne, and Sigman challenge the functional ‘fixity’ or ‘givenness’ of the material environment, and thus highlight the transformative potential of materiality in human life. The artistic practices of Group Material, Ai, and van Heeswijk demonstrated the potentials of, and challenges for, socially engaged art to mobilize communities in this transformatory process by attempting to enable residents and users to take ownership of and directly engage with the complex structures of their urban environment. The importance of acknowledging and working with the entanglement between the artistic practice and its social and material site was here foregrounded. This entanglement points to how mobilizing communities in socially engaged art is not solely about representing various perspectives, but also depends on the settings of participation and its representative framework in terms of what kinds of perspectives can be enabled and represented.

In Part III, by turning attention to the project *invisible Zürichs* and my own artistic practice with zURBS, I provided a critical and self-reflexive discussion about the settings of participation that are produced through socially engaged art. The contradictions in participatory processes that attempt to enable residents to engage in their material environments were here foregrounded. While participatory techniques are often deployed to simply help people accept the structural conditions of their lives, *invisible Zürichs* illustrated how socially engaged art may provide an alternative participatory process that enables participants to rather transgress these conditions. Focusing on personal narratives and engagements with urban space can here be seen as a way of countering dominant and standardized uses of space, which at the same time produce the city not as a spatial framework external to its users, but as produced through them and productive of them.

However, the participatory process within *invisible Zürichs* was also ridden with challenges posed by contradictions between instrumental and emancipatory participation: our attempts to defamiliarize the everyday in order to 'change the perception' resulted in the fact that many participants could not see the link to and relevance for their everyday lives; the liberatory potential that we assigned to our aesthetic framework was too abstract or restricting for participants who were unfamiliar with these kinds of events; our aesthetic approach favoured a certain form of engagement that contradicted the very openness we were trying to create, and at the same time total openness entailed the insecurities of not knowing what to do, and thus had to be countered by giving the participants concrete guidelines and directions that risked being perceived as too directional. These contradictions illustrated the complex and contested setting for participation in socially engaged art and raised further questions about the representation of people and their interests.

In Part IV, I scrutinized the contradictory field of representation in relation to zURBS' projects *stadtARCHIV, Montopia*, and *St. Clement's Utopolis*. I analysed what kinds of participation were enabled within the particular frames of representation provided by each project. I here illustrated how representation is tenuous, unstable, and difficult to achieve convincingly, and pointed to how the zURBS team struggled with keeping our artistic aims intact, while at the same time working within various institutional constraints and expectations. *stadtARCHIV* illustrated how 'the logic of the institution' bounded the possibilities for agency and inclusion in terms of the subject-positions made available in the archive we produced. Even though we moved the stadtARCHIV into the streets in order to escape this logic, it did not change the cultural field that defines the limits of collaboration. While being wary of not alienating passers-by and enforcing a distinction between those who are inside and those who are outside of the art discourse, we focused so much on openness and agreement that we risked stereotyping the participants by not sufficiently engaging with their views and opinions. *Montopia* illustrated another challenge in this regard. We wanted the workshops to operate on a level removed from the everyday in order to 'free' the participants from their normal routines and habits. However, everyday prejudices relating to the participants' social situations and identities in Monthey led to antagonistic encounters that reinforced the binary camps of 'us and them'. In *St. Clement's Utopolis*, we – the zURBS team – enforced similar binary camps, only this time between us, as workshop facilitators, and the participants. Instead of focusing on how the workshop could engender new subjectivities, we concentrated on the self-expression of the preconceived participant identities that were communicated to us by our institutional collaborators.

stadtARCHIV, *Montopia*, and *St. Clement's Utopolis* illustrate several contradictions within issues of representation. These contradictions relate to how the 'invited space' of our workshops affected the ability of participants to enter and exercise voice on their own terms on several levels. First, the 'safe' frame we aimed to provide for the participants in terms of taking into account specific needs and abilities risked essentializing pregiven identities rather than producing new ones. Second, fostering a personal engagement based on subjective perceptions, positions, and understandings could ultimately compromise the creation of a common ground and collective ownership where people can address each other as co-producers of the city. Third, antagonistic encounters between 'us and them' compromised collaborative inventions and a sense of collective responsibility. However, focusing excessively on harmonious reconciliation risked stereotyping the participants. And finally, artistic authorship was needed in order to provide a form of transformative experience but could place us in the authoritarian role of the privileged bearers of insight. These contradictions elucidated the limitations of socially engaged art in terms of aiming to provide a so-called 'free' space from which alternatives to the present can be imagined and articulated.

Working with contradictions

Vanesa Broto (2015) observes that the diagnostic mechanism to contradictions is always the same: 'contradiction points towards an impossibility that needs to be resolved through the annihilation of one of the terms of the contradiction' (p. 461). As I will discuss later in this Conclusion, Broto advocates a more dialectical notion of contradiction that brings forth its generative potential. This dialectical approach chimes with Harvey's (1996) conception of 'dialectical thinking' as a way of overcoming static conceptions of contradictions by questioning the processes by which these contradictions are constituted. However, this dialectical view is often neglected in practice. As Broto points out, there is a tendency for actors seeking social change to attempt to resolve contradiction by countering static opposites:

passive residents should be turned into active participants, homogeneous urban space is to be countered with a heterogeneous and liberatory space, alienation is to be exchanged for intimacy and personal engagement, individualization should be replaced with collectivity, and so on. In order for these contradictory forces to cancel each other out (i.e. in order for passive to become active, for homogeneous to become heterogeneous), there is a risk, as I have illustrated throughout this research, to posit alternatives as a liberatory exteriority to the everyday sphere of production of meaning, knowledge, discourse, and institutions. This liberatory exteriority constitutes the aims of producing collectivity, activation, and heterogeneity against an outside (individuality, passivity, and homogeneity) that is seen as external and antagonistic to these aims.

However, as Joseph (2002) points out, referencing Butler, oppositional agency is a reiterative or rearticulatory practice that is immanent in power, and not a relation of external opposition to power. In other words, power is productive of subjectivity, not simply repressive or exploitative in relation to individual subjects. Technocratic and neoliberal institutions and practices are, for example, to some extent the outcome of some element of desire or need for these institutions. Hence, Purcell (2013) argues that people have the desire to be ruled, 'to be relieved of the burden of ruling ourselves' (p. 93). For example, in the contemporary Global North, according to Purcell, the desire to be ruled manifests itself as a taste for the subtle and seductive oligarchy of consumer capitalism turning us into willing subjects of inertia, inactivity, and passivity. The project of (urban) democracy, then, is not so much to confront a power wielded by benevolent forces beyond our control as it is about reappropriating our own power by reactivating our desire for democracy. As discussed in Part I, democracy here makes space for political action in which we seek to lay new grounds by extrapolating and amplifying practices and ideas that are already taking place. Accordingly, liberatory social change cannot come from some external non-rationalized realm of emergent truth and freedom, but only from the constraints of the present.

As this book illustrates, the spaces we created in our workshops, the settings for meaning-making, participation, and representation, were the product of social and material circumstances, and not their exteriority. The initial researcher costumes we wore reminded participants of Greenpeace activists, the workshops' location within cultural institutions compromised their 'openness' for people who were not familiar with this setting (such as the couple in Zurich), and also bounded the possibilities for agency in terms of what subject-positions were available for protest movements such as Wagenplatz in Basel. I have shown that socially engaged art alas cannot break through the discipline of everyday neoliberal production and free political subjects from structures of power. However, this form of artistic practice does enable a re-enactment of the multiple articulations that determine these structures, and this way creates contexts in which critical reflection, questioning, and change are made available. This was illustrated through situations in which, for example, the female student confronted her insecurity about walking through an empty park; the group that encountered the friendly and culinary homeless man with the thistles, who countered the negative image of homeless people as disturbing the harmony of the city; the negotiation of 'messy', 'unclean', and 'unorderly' details of urban space as materialized relationships of proximity that nourished residents' collective identity and sense of place, rather than being seen as disturbances and 'matter out of place'; the school kids realizing, by attending to Monthey in new ways, that littering is not something they desire in their city; the teenagers discovering that they indeed did have an opinion about the city, articulating the need for more 'free spaces' in which they could hang out; and so on.

Although structural issues relating to, for example, housing and finance were not explicitly raised in these everyday critical reflections, the situations problematized naturalized understandings of city space, and thereby functioned as a first step towards negotiating better urban futures. Important here is that the situations show how socially engaged art can be productive of existing material and institutional conditions and articulate alternatives through participation in them, rather than escaping from or ranging freely outside them. From this non-binary perspective, it is possible to look beyond the opposition of contraries such as outside/inside, active/passive, and recognize socially engaged art as a co-production between various actors through different levels of involvement. For example, in *stadtARCHIV* and *invisible Zürichs*, we involved the employees in various hotel receptions, kiosks, shops, cafés, and bars to help hand out envelopes to the participants. The encounters between the participants and these helpers were not ones of commercialized and controlled activity based on economic exchange (with associated 'demands' of passivity). Rather than being based on monetary transaction, the exchange was based on a personal interaction in which stories and views on the city were shared. The opposition between participant and non-participant, activity and passivity, was here blurred as perceived 'outsiders' revealed themselves as being part of the workshop. As mentioned in Part III, this often came as a surprise to participants, who did not expect 'the woman in the sex shop', 'the guy in the video store', or 'the kebab seller' to be involved in the workshops.

For Rancière (2009), this is what emancipation means: the verification of equality by blurring the opposition between those who look and those who act. This definition challenges the idea of 'becoming active' as implying that people have been dispossessed of their rightful activeness, which needs to be restored. Instead, the non-binary perspective suggests that we start by recognizing the plural appearances of participation. In addition to the 'helpers' in the shops, bars, cafés, and reception desks, the workshop participants engaged with various people that they encountered and approached during their urban exploration, such as the group that decided to share their collective story about the man sitting everyday waiting for an adventure with the main character himself, or the groups that encountered the homeless man, as mentioned above. These people became part of the workshops by having their stories (re)told and archived by the participants. Additionally, some workshop participants reported that they had been approached by security guards and private persons wanting to know why they were suspiciously lurking around/loitering in back alleys and side streets with no obvious aim or sense of direction. It was clear that the groups' presence in urban space was felt and observed by other users as a form of transgression from the 'proper' uses of that space, as discussed in Part III. These individuals engaged with the workshop participants on the grounds of personal interest and curiosity, and in this way learned about the project. Consequently, some took part in the workshops by helping to look for the envelopes, or by giving tips to nearby places that could be worth checking out for their imaginative potential. These interactions blurred the 'frame' of the workshop along with its participatory identities. In our projects in Zurich and Basel, this was manifested in situations such as those above, where the participants were no longer sure who (or what) was part of the workshop or not. In Monthey, this aspect was made explicit in the exhibition of *Montopia*, where it was unclear which participants had made which models. In turn, these situations opened to scrutiny fixed subject-positions and given representative frames.

Generating a desire for change

As illustrated above, socially engaged art can work as an active and generative force in questioning given 'truths' and identities. This questioning relates to an awareness of contradictions between what is (re)presented to us as real and what we experience as being real. The philosopher Alexandre Kojève (1969, cited in Broto 2015, 462) here argues that 'to become aware of a contradiction is necessarily to want to remove it. Now, one can in fact overcome contradiction of a given existence only by modifying the given existence, by transforming it through actions'. As Broto (2015) points out, contradictions can be seen as activating a desire for intervention, and thus provide a perspective that enables a diagnosis of instances of concrete reality. The moments in which our workshops worked most productively with contradictions in this regard were when the space produced through the workshops and urban space appeared as equally real and constructed. This is perhaps best illustrated in the *Montopia* exhibition that undermined the discourse of 'reality' by usurping it, and hence foregrounded the way that reality and representation mutually imply each other. There were also moments in other projects, when the participants were out in the city, when the imaginative layer of the workshop and 'the landscape of the real' overlapped and blurred. For example, in the *invisible Zürichs* workshops, when one group discovered what they thought was a sound installation in a phone booth, not knowing if we had created it or not, or when another group was convinced that we had placed all the objects that they found, or when a group looked for an envelope where the clue mentioned a woman in a red dress and suddenly a woman in a red dress walked swiftly passed them.

One of the participants referred to the film *The Truman Show* (1998) when describing how she experienced the blurring of 'fiction' and 'reality' in the workshops. This reference is interesting because the film illustrates how the contradiction between imaginary and actual space offers an opening to an emancipatory urban space (Pratt and San Juan 2004). Truman lives happily in the prototype town of 'new urbanism', Seaside, located on a utopic island in Florida. He is completely unaware that his life is the focus of a reality TV show aired since his birth, that the town is a giant set piece encased in a climate-controlled dome, and that everyone around him is an actor going by a script. It is when Truman becomes aware of the contradictions within his controlled life that it generates a desire to break his routines and explore a life beyond Seaside: at some point, he deviates from the route of his everyday walk and discovers the 'backstage' area where actors are having a break. Truman's discovery of these contradictions makes him realize the limits of his environment and at the same time opens up new possibilities as he embarks on a journey to overcome both the given truth (life in Seaside as he knows it) and its opposition (the constructed Seaside that is revealed to him). Geraldine Pratt and Rose Marie San Juan (2004) point out that it is not a coincidence that the decisive site of the film is a transitional space between inside and outside. Based on contradiction, this site becomes a horizon that simultaneously delimits Truman's world and opens up a new space.

Emancipation is not depicted here as a concrete urban space or form. Rather, it is seen as a process that implies the capacity to leave particular spaces: Truman finds his path by means of and in relation to the concrete urban site of Seaside. Hence, as Pratt (2004) argues, seeing emancipation as a process is not the same as ignoring its material referent. Instead, it opens up transformatory possibilities in concrete experiences of specific places and societies. It is, as Jameson (2009) points out, 'the detective work of a decipherment and a reading of Utopian clues and traces in the landscape of the real' (p. 415). From this perspective, there is no 'outside' from which resistance or opposition can be articulated. Change is only enabled in relation to present structures and systems rather than as an escape from them.

Harvey (2000) fears that the focus on emancipation as a process provides no concrete direction for how this change may come about. However, as I have illustrated through my practice, overcoming contradictions requires explaining the conditions that produce the contradiction, and may this way provide a direction towards broader reconfigurations of social and material practice (Broto 2015). For example, in Zurich, the contradiction between the official promotion of a clean and orderly city and the participants' celebration of the unclean and unorderly aspects of urban space brought into question the exclusionary socio-spatial orders produced by the idea of 'cleanliness' in relation to a sense of being 'in place' or 'out of place'. Accordingly, the urban planner and participant Roger expressed a desire to organize similar workshops with investors to make them 'think about this micro-cosmos that they are destroying' in many new urban (re)developments in Zurich. In Monthey, the contradictory views of *Montopia* as mere fantastical exaggeration and *Montopia* as possible vision for Monthey raised questions around agency and authority in relation to the future of the city. Marco, the theatre director, accordingly expressed a desire to create a 'citizen lobby' promoting *Montopian* visions to local authorities in order to stir a debate about what is possible and what is not possible for a perceived small city such as Monthey. Furthermore, the schoolteachers in Zurich, Basel, London, and Monthey, comparing zURBS' workshops with their normal teaching methods, all expressed a desire to incorporate some of zURBS' approaches (storytelling, model-making, urban exploration) in their day-to-day teaching.

These examples demonstrate not just that socially engaged art may inspire changes, but also that such changes are embedded in wider totalities beyond the scope of the practice itself. In this view, then, the key concern of our workshops was not to provide concrete alternatives and ways of overcoming contradictions, but to use them to raise critical questions about the spatial and material organization of the urban environment. In the context of socially engaged art, contradictions here point to an overall direction rather than a prescribed pathway.

Towards socially engaged art as a mosaic of contradictions and possibilities

> I continue to believe in the appeal to hunger, in the classes that are hungry. If I were a specialist in food production [. . .], I would devote myself to issues concerning how to feed millions of people, which implies changes to the most stubborn of cultural habits [. . .] But, instead, I am a specialist in imaginative and verbal material, and I dedicate myself to the hunger for written words, for stories told, for mythological figures: all stuff that is no less essential than food, as we all know.
>
> (Italo Calvino, cited in Modena 2011, 17)

After spending the week together in the Swiss mountains, Sabeth, Nina, and I had decided on zURBS' future: we unanimously wanted zURBS to continue. Our sometimes-fierce disagreements, our constant doubts and frustrations, our failures and struggles bore witness to something too good to let go: a deep passion for, understanding of, and belief in the potential of the arts to raise individual and collective reflection on the state of urban realities and generate desires for change. Throughout our various projects, we had, like Calvino, become specialists in imaginative material and storytelling, turning urban space upside down and inside out through our methods and approaches. We had trained ourselves and our participants in transduction, to seek out the possibilities contained in the present, and we had a responsibility to keep these possibilities alive through our practice rather than drowning them in our own insecurities. However, it was clear to us that in order to keep going, we had to reconfigure our practice according to some of the issues I have discussed in this conclusion.

I have here argued for the potential of positioning contradiction as a starting point for socially engaged art, rather than as some kind of negation of it. I have illustrated how contradiction can act as a ground for experimentation and disciplinary openness by providing non-binary perspectives and desires for change. Contradiction links art and the social, aesthetics and ethics, through action that opposes false clarity and simplicity in political discourse to the conflictual, ambivalent, intersubjective space of critical transformative experience. Hence, it is part of the material reality of socially engaged art. As I have illustrated in this book, where contradiction is not made productive as a vital element of socially engaged art, even the most rigorous ethical considerations or carefully guarded artistic autonomy becomes an extension, rather than a reconfiguration, of existing power structures and what is presented to us as given. This argument may be further illustrated with an example borrowed from Dikeç (2012). Think of a jigsaw puzzle and a mosaic as two different 'models' for socially engaged art. In the puzzle, pieces can be physically moved around, but they only fit in one way. There is only one rational and predetermined way of assembling the pieces. In the mosaic, on the other hand, it is possible to imagine a variety of assemblages without even moving the pieces around. The individual pieces are in the mosaic, but the final outline is not given. Depending on the outline you imagine, different pieces will be related to each other, producing different forms and making new connections each time.

Socially engaged art is stuck in the jigsaw puzzle when it is considered to be inherently liberatory, communitarian, or political. This position depends on splitting the world into two levels: the false and oppressive representative hegemony of everyday life, and the good and pure reality revealed through socially engaged art. Rather than mediating between urban imaginaries and the material environment, this approach foregrounds their mutual incompatibility.

The model of the mosaic starts from a different position: socially engaged art is not an alternative to everyday life, nor can it provide alternative urban imaginaries. Rather, this practice is, along with everyday life, an equally integral component of the larger system called our lifeworld. The interdependence of urban imaginaries and the material environment is here foregrounded. This interdependence focuses on contradictions and limits, and moves the debate away from the stalemate produced by stubborn opposition in order to shift attention to new avenues where productive change can be brought about.

Looking at our own practice through the model of the mosaic enabled us to stake out three 'new avenues' for change as a product of the internal contradictions within zURBS. First of all, we had to rethink our approach to participation. The question of participation had caused fierce disagreement and discussion throughout our work, with Sabeth and I being in stark opposition. The way I understood it, she wanted to move away from participation altogether, and just do non-collaborative projects on our own. She, on the other hand, regarded my position as one advocating participation for its own sake, regardless of the ability or interests of people to participate. The idea of the plural appearances of participation enabled us to transgress this opposition and rather reconfigure the participatory elements of our projects. We agreed that participation was not just about having people 'do' our workshops, but opening up a field in which people could be involved in different ways, be it as workshop participants, helpers, or simply observers. During his artistic practice in public space (e.g. *Bataille Monument* in Kassel, 2002), Thomas Hirschhorn (cited in Riet et al. 2012, 143) witnessed how people often preferred to observe the intervention from a distance. Eventually, some would actively take part, but others would stay in this 'second row' sphere. Hirschhorn learned to respect this choice and not to press people to come and join. He explained that the 'second row' is as much part of the project as the people that actively participate, as the presence of the intervention in the existing urban context is still noticed and comprehended.

Second, we had to redefine our own roles and the collaborative element in our projects. Up until now (with the exception of 'the mobile *stadtARCHIV*') our participation in the workshops had been solely as that of facilitators. We provided an imaginative and performative framework for the participants to explore, but we did not collaborate with the participants in the exploration as such. Apart from 'setting the scene' and mediating the storytelling that would take place in the workshop process, we were mere observers, sending the participants out into the city and waiting anxiously for their return. With 'the mobile *stadtARCHIV*', the situation was different. In addition to engaging with passers-by, Nina, Sabeth, and I were busy archiving on our own, collecting objects, taking notes, making drawings, and so on. Sabeth admitted that she felt much more confident in a project when she was busy genuinely exploring the same tasks and questions that she wanted the participants to explore. By adopting this more active participatory approach, we wanted our future projects to be a more explicitly collective effort towards some common aim, rather than us 'giving' the participants a particular experience that we had carefully prepared for them.

Third, we decided that from now on, zURBS would be called 'urban performance collective' and our workshops would be called 'urban performances'. This way, we would hopefully better avoid situations in which we had to downplay the performative and playful elements of our projects to a degree where it compromised the participants' experiences. Who had ever heard of a performance company being asked to remove their scenography or skip their costumes out of concern for their audience? Whereas the word 'workshop' signals a somewhat clearly defined space, such as a form of educational seminar where people engage in a study or work on a creative project or subject, the words 'urban performance' were meant to signal the constructedness of the world and zURBS practice alike, thus blurring the frames of the experience as something that is part of everyday life rather than external to it. This way, our 'urban performances' suggested the transmutability of things, training participants to search for alternatives within the present – to take apart the urban environment and combine or rearrange elements to form new social and material possibilities.

This transformational and combinatorial approach to reality is well illustrated in Calvino's *Invisible Cities*. It seems pertinent to end here by returning to this text, which has been a faithful companion throughout my research and practice. The idea of the mosaic as a sphere of multiplicity, sampling possibilities and exploring potentials, is illustrated in the city of Fedora. The museum in Fedora contains a crystal globe in every room, and in every globe 'you see a blue city, the model of a different Fedora [. . .], the forms the city could have taken' (Calvino 1979, 28). The inhabitants of Fedora visit the museum and project their desires onto their most meaningful models, imagining what sensory experience they would have felt had they inhabited one of those ideal cities. For the first time in the book, Polo interrupts himself in the middle of the description of the city and addresses Khan somewhat didactically:

> On the map of your empire, O Great Khan, there must be room for the big stone Fedora and the little Fedoras in the glass globes. Not because they are all equally real, but because all are only assumptions.
>
> (Calvino 1979, 28)

Polo here confers particular relevance on the idea that the material world itself is merely an assumption, and nothing in the political or social order is a historical necessity. As the late novelist, semiotician, and philosopher Umberto Eco, one of *Invisible Cities*' most faithful readers, put it:

> The presence of possible Fedoras that may never exist as such allows us to think that the Fedora of the present is perhaps not so necessary and may therefore be changed. We are not taken back in time to change the possibility that materialized. Rather, by contemplating the counterfactual in which its opposite materializes, by leaping backwards playfully we dash forward for real, in search of a third possibility not yet given, but whose possibility has been revealed through the game of nostalgic combinatory of the possible.
> (cited in Modena 2011, 179)

Bibliography

Broto, V. C. 2015. "Contradiction, Intervention, and Urban Low Carbon Transitions." *Environment and Planning D: Society and Space* 33 (3): 460–476.

Calvino, I. 1979. *Invisible Cities*, 2nd edition. London: Pan Books.

Dikeç, M. 2012. "Space as a Mode of Political Thinking." *Geoforum* 43 (4): 669–676.

Harvey, D. 1996. *Justice, Nature and the Geography of Difference*. Oxford: Blackwell.

———. 2000. *Spaces of Hope*. Edinburgh: Edinburgh University Press.

Jameson, F. 2009. *Valences of the Dialectic*. London: Verso.

Joseph, M. 2002. *Against the Romance of Community*. Minneapolis, MN: University of Minnesota Press.

Kojève, A. 1969. *Introduction to the Reading of Hegel: Lectures on the Phenomenology of Spirit (Assembled by Raymond Queneau)*, edited by A. Bllom. New York: Basic Books.

Modena, L. 2011. *Italo Calvino's Architecture of Lightness*. New York: Routledge.

Pratt, A. C. 2004. "The Cultural Economy: A Call for Spatialized 'Production of Culture' Perspectives." *International Journal of Cultural Studies* 7 (1): 117–128.

Pratt, G. and San Juan, R. M. 2004. "In Search of the Horizon: Utopia and The Truman Show and The Matrix." In *The Emancipatory City? Paradoxes and Possibilities*, edited by L. Lees, 192–209. London: Sage.

Purcell, M. 2013. *The Down-Deep Delight of Democracy*. Chichester: John Wiley & Sons.

Rancière, J. 2009. *The Emancipated Spectator*. London: Verso.

Riet, S., van Eeghem, E., Verschelden, G., and Dekeyrel, C. 2012. *Reading Urban Cracks: Practices of Artitsts and Community Workers*. Ghent: MER. Paperkunsthalle.

Index

Note: Page numbers in *italic* refer to figures.